有色行业大学英语系列教材

冶金工程英语

主　编　汪婷婷
副主编　邓　燕

·长沙·

图书在版编目(CIP)数据

冶金工程英语 / 汪婷婷主编. —长沙:中南大学出版社,2021.8

有色行业大学英语系列教材

ISBN 978-7-5487-4484-9

Ⅰ. ①冶… Ⅱ. ①汪… Ⅲ. ①冶金－英语－高等学校－教材 Ⅳ. ①TF

中国版本图书馆 CIP 数据核字(2021)第 112733 号

冶金工程英语
YEJIN GONGCHENG YINGYU

主　编　汪婷婷
副主编　邓　燕

□责任编辑	刘小沛			
□责任印制	唐　曦			
□出版发行	中南大学出版社			
	社址:长沙市麓山南路		邮编:410083	
	发行科电话:0731-88876770		传真:0731-88710482	
□印　　装	长沙印通印刷有限公司			
□开　　本	787 mm×1092 mm　1/16	□印张 9.5	□字数 273 千字	
□版　　次	2021 年 8 月第 1 版	□印次 2021 年 8 月第 1 次印刷		
□书　　号	ISBN 978-7-5487-4484-9			
□定　　价	40.00 元			

图书出现印装问题,请与经销商调换

Foreword

在新工科和新文科建设持续推动的学科交叉融合发展的大趋势之下，大学英语教学应将满足国家战略发展需求、社会需求、时代需求以及学生个性发展需求，为国家战略需求、学术前沿研究、学生职场需求等服务并提供语言技能与研究方法上的支持作为初心和使命。为此，江西理工大学全面贯彻教育部《大学英语教学指南（2020版）》所强调的"注重人文性和工具性的统一"要求，主动对标国家、社会、时代、学科发展、学生个性发展之需求，在大学英语教学改革中为实现国家、时代、社会需求与学生个性需求相统一、人文性与工具性相统一、隐性教学目标与显性教学目标相统一的"三个统一"目标，组织策划编写了"有色行业大学英语系列教材"，包括《新文科专业英语》《采矿工程英语》《土木工程英语》《测绘工程英语》《电子信息工程英语》《电气工程英语》《冶金工程英语》《材料工程英语》和《机电工程英语》共9部教材。

本套教材有以下特色：

跨学科交融。教材内容丰富，范围广泛，涉及采矿、土木、材料、冶金、机电、电气、信息、测绘、经济、教育和认知科学等自然科学和社会科学内容，跨学科性强，同时自然科学在教材内容中的比重得到提高和加强；专门用途英语等跨学科英语教学进一步拓展了教材内容的跨学科空间。外语学科与专业学科领域的融合，有利于提高学生英语学习的积极性，整合其专业知识与语言知识，提高学生运用行业英语的综合能力。

选材去同质化。本套教材实现了 EGP 和 ESP 的有机结合，较好地解决了传统 ESP 教学面临的教师专业知识缺乏以及教学可行性较差的问题。选材具有通用性、恰当性。在语体上接近实际交际语言，在呈现方式与练习形式上体现了英语学习的实践性，在教学方法上体现了可操作性；教材教学设计充分利用计算机、网络、多媒体等现代技术，打破传统教材单一文字的形式，结合音频、图像、视频等输入方式，使其立体化、多方位、多层次地配合教学。教材教学目标合理可及，实现了与专业英语教学的有效衔接，为学生后续的专业英语学习做好词汇、知识和技能等方面的储备，如理解专业常见英语术语以及一般学术词汇，掌握专业领域基本概念、本专业文献英语语篇表达方式，听懂相关英语学术讲座及读懂相关文献资料等。

弘扬中华优秀传统文化。教材结合落实立德树人的根本任务，在教学内容的选择、讨论话题的设计方面，着力提升学生正确的理想信念，强化学生的政治认同、家国情怀和文化素养，提高学生的民族自豪感，增强学生宣传中华优秀传统文化的能力。

教材的编写得到了江西理工大学各级领导的高度重视和支持。副校长龚姚腾教授亲自审定编写大纲和推动教材的编写，教务处处长吴彩斌教授为教材的编写提供了宝贵的意见，并纳入学校整体教材编写规划，为教材的出版提供了资金支持。

教材自2019年5月酝酿启动以来，历时两年，由外国语学院牵头，稀土学院、资源与环境工程学院、土木与测绘工程学院、建筑与设计学院、材料冶金化学学部、机电工程学院、电气工程与自动化学院、信息工程学院、经济管理学院、法学院、应急管理学院、马克思主义学院等共同参与了编写和审校，在此一并表示感谢。

限于时间和水平，本系列教材如有不当之处，敬请各位同行批评指正，以便及时修改完善。

教育部高等学校大学外语教学指导委员会委员
江西理工大学外国语学院院长
邓晓宇教授
2021年秋

Contents

Unit 1　Introduction ··· 1

Unit 2　Iron Making ·· 23

Unit 3　Steel Making ·· 45

Unit 4　Introduction to Nonferrous Metallurgy ··· 64

Unit 5　Powder Metallurgy ··· 85

Unit 6　Development of Nonferrous Metallurgy ··· 107

Glossary ·· 130

Unit 1　Introduction

Part Ⅰ：Text A

Section A　Warm-up Questions

Discuss with your partner about the following questions.
1. How much do you know about the history of iron and steel?
2. What is pig iron?
3. How did pig iron get this name?

Section B　Listening Practice

Listen to the passage and fill in the blanks below.

Our modern world, with its skyscrapers, suspension bridges, and automobiles, relies heavily on steel. If you own an 1._____ car, however, you know first hand that steel has one 2._____ problem: it rusts. All steel, that is, except stainless steel. What keeps this remarkable material, used commonly in kitchen sinks and cookware, from rusting?

Steel's main 3._____ is iron. When steel rusts, its iron combines with oxygen and reverts to iron ore, the raw state from which it came. As this happens, the steel turns brown and begins to crumble. There are many ways to protect steel from corrosion. Steel is sometimes 4._____ or greased, or coated with a metal that is less likely to rust.

Sometimes a metal that is more likely to rust is attached. Known as a sacrificial metal, this works by drawing the 5._____ process away from the steel. For example, bars of zinc are attached to the hulls of some ships. The zinc rusts heavily, but the steel hull stays 6._____ safe.

The best way to avoid rust is to use stainless steel. Like all steel, stainless steel is mostly iron, but it also contains nickel and chromium. These are not just a protective coating, but are 7._____ into the steel itself. The mixture must contain at least ten percent chromium, because it's the chromium that protects stainless steel from corrosion.

What happens is this: Like a sacrificial metal, the chromium rusts first. Unlike iron, however, rusting chromium doesn't crumble 8._____. Instead, it forms an invisibly thin layer that protects the iron underneath. The nickel in stainless steel helps hold this protective layer of chromium rust in place. Remember that chromium and nickel are present 9._____ stainless steel, not just on the surface. Because of this, the microscopic layer will form itself anew, even when the steel is cut or 10._____.

☞ Section C Active Reading

The History of Steel

1 The development of steel can be traced back 4000 years to the beginning of the Iron Age. Proving to be harder and stronger than bronze, which had previously been the most widely used metal, iron began to displace bronze in **weaponry** and tools. For the following few thousand years, however, the quality of iron produced would depend as much on the **ore** available as on the production methods.

2 By the 17th century, iron's properties were well understood, but increasing urbanization in Europe demanded a more **versatile** structural metal. And by the 19th century, the amount of iron being consumed by expanding railroads provided **metallurgists** with the financial **incentive** to find a solution to iron's **brittleness** and inefficient production processes.

3 Undoubtedly, though, the most important breakthrough in steel history came in 1856 when Henry Bessemer developed an effective way to use oxygen to reduce the carbon content in iron: The modern steel industry was born.

4 At very high temperatures, iron begins to absorb carbon, which lowers the **melting point** of the metal, resulting in **cast iron** (2.5% to 4.5% carbon). The development of **blast furnaces**, first used by the Chinese in the 6th century BC but more widely used in Europe during the Middle Ages, increased the production of cast iron.

5 **Pig iron** is **molten** iron run out of the blast furnaces and cooled in the main channel and **adjoining molds**. The large, central and adjoining smaller **ingots** resembled a sow and suckling **piglets**. Cast iron is strong but suffers from brittleness due to its carbon content, making it less than

ideal for working and shaping. As metallurgists became aware that the high carbon content in iron was central to the problem of brittleness, they experimented with new methods for reducing the carbon content to make iron more workable.

6　By the late 18th century, ironmakers learned how to transform cast pig iron into a low-carbon content wrought iron using **puddling** furnaces (developed by Henry Cort in 1784). The furnaces heated molten iron, which had to be stirred by puddlers using long, oar-shaped tools, allowing oxygen to combine with and slowly remove carbon. As the carbon content decreases, iron's melting point increases, so masses of iron would **agglomerate** in the furnace. These masses would be removed and worked with a forge hammer by the puddler before being rolled into sheets or rails. By 1860, there were over 3000 puddling furnaces in Britain, but the process remained hindered by its labor and fuel intensiveness.

7　One of the earliest forms of steel, **blister steel**, began production in Germany and England in the 17th century and was produced by increasing the carbon content in molten pig iron using a process known as **cementation**. In this process, bars of wrought iron were layered with powdered **charcoal** in stone boxes and heated. After about a week, the iron would absorb the carbon in the charcoal. Repeated heating would distribute carbon more evenly and the result, after cooling, was blister steel. The higher carbon content made blister steel much more workable than pig iron, allowing it to be pressed or rolled.

8　Blister steel production advanced in the 1740s when English clockmaker Benjamin Huntsman while trying to develop high-quality steel for his clock springs, found that the metal could be melted in clay **crucibles** and refined with a special flux to remove slag that the cementation process left behind. The result was a crucible, or cast, steel. But due to the cost of production, both blister and cast steel were only ever used in specialty applications. As a result, cast iron made in puddling furnaces remained the primary structural metal in industrializing Britain during most of the 19th century.

9　The growth of railroads during the 19th century in both Europe and America put enormous pressure on the iron industry, which still struggled with inefficient production processes. Steel was still unproven as a structural metal and production of the product was slow and costly. That was until 1856 when Henry Bessemer came up with a more effective way to introduce oxygen into molten iron to reduce the carbon content.

10　Now known as the Bessemer Process, Bessemer designed a pear-shaped receptacle, referred to as a "converter" in which iron could be heated while oxygen could be blown through the molten metal. As oxygen passed through the molten metal, it would react with the carbon, releasing carbon dioxide and producing a more pure iron.

11 The process was fast and inexpensive, removing carbon and silicon from iron in a matter of minutes but suffered from being too successful. Too much carbon was removed, and too much oxygen remained in the final product. Bessemer ultimately had to repay his investors until he could find a method to increase the carbon content and remove the unwanted oxygen.

12 At about the same time, British metallurgist Robert Mushet acquired and began testing a **compound** of iron, carbon, and **manganese**, known as **spiegeleisen**. Manganese was known to remove oxygen from molten iron and the carbon content in the spiegeleisen, if added in the right quantities, would provide the solution to Bessemer's problems. Bessemer began adding it to his conversion process with great success. One problem remained. Bessemer had failed to find a way to remove **phosphorus**, a **deleterious** impurity that makes steel brittle, from his end product. Consequently, only phosphorus-free ore from Sweden and Wales could be used.

13 In 1876 Welshman Sidney Gilchrist Thomas[1] came up with the solution by adding a chemically basic flux, **limestone**, to the Bessemer process. The limestone drew phosphorus from the pig iron into the slag, allowing the unwanted element to be removed. This innovation meant that, finally, iron ore from anywhere in the world could be used to make steel. Not surprisingly, steel production costs began decreasing significantly. Prices for steel rail dropped more than 80% between 1867 and 1884, as a result of the new steel producing techniques, initiating the growth of the world steel industry.

14 In the 1860s, German engineer Karl Wilhelm Siemens further enhanced steel production through his creation of the **open-hearth** process. The open-hearth process produced steel from pig iron in large shallow furnaces. The process, using high temperatures to burn off excess carbon and other impurities, relied on heated brick chambers below the hearth. Regenerative furnaces later used exhaust gasses from the furnace to maintain high temperatures in the brick chambers below.

15 This method allowed for the production of much larger quantities (50~100 metric tons could be produced in one furnace), periodic testing of the molten steel so that it could be made to meet particular specifications and the use of scrap steel as a raw material. Although the process itself was much slower, by 1900, the open-hearth process had primarily replaced the Bessemer process.

16 The revolution in steel production that provided cheaper, higher quality material, was recognized by many businessmen of the day as an investment opportunity. Capitalists of the late 19th century, including Andrew Carnegie and Charles Schwab, invested and made millions in the steel industry. Carnegie's US Steel Corporation, founded in 1901, was the first corporation ever launched valued at over one billion dollars.

Unit 1 Introduction

◆ Words and Expressions

weaponry /ˈwepənri/ *n.* all the weapons of a particular type or belonging to a particular country or group 武器，兵器

ore /ɔː(r)/ *n.* rock, earth, etc. from which metal can be obtained 矿石

versatile /ˈvɜːsətaɪl/ *adj.* (of food, a building, etc.) having many different uses 多用途的；多功能的

incentive /ɪnˈsentɪv/ *n.* something that encourages you to do something 激励；刺激

brittleness /ˈbrɪtlnəs/ *n.* the fact of being hard but easily broken 脆性

adjoining /əˈdʒɔɪnɪŋ/ *adj.* next to or joined to something 邻接的；毗连的

mold /məʊld/ *n.* a container that you pour a liquid or soft substance into, which then becomes solid in the same shape as the container 模具；铸模

piglet /ˈpɪɡlət/ *n.* a young pig 猪仔；小猪

puddle /ˈpʌdl/ *n.* a small amount of water or other liquid, especially rain, that has collected in one place on the ground 水洼；小水坑

agglomerate /əˈɡlɒməreɪt/ *v.* to form into a mass or group; to collect things and form them into a mass or group (使)成团，聚结

compound /ˈkɒmpaʊnd/ *n.* a thing consisting of two or more separate things combined together 复合物；混合物

deleterious /ˌdeliˈtɪəriəs/ *adj.* (formal) harmful and damaging 有害的；造成伤害的；损害的

◆ Terminology

metallurgist /məˈtælədʒɪst/ *n.* 冶金学家

melting point /ˈmeltɪŋ pɔɪnt/ 熔点

cast iron /ˌkɑːst ˈaɪən/ *n.* 铸铁

blast furnace /ˈblɑːst fɜːnɪs/ *n.* 鼓风炉；高炉

pig iron /ˈpɪɡ aɪən/ *n.* 生铁；铸铁

molten /ˈməʊltən/ *adj.* (金属、岩石或玻璃)熔化的；熔融的

ingot /ˈɪŋɡət/ *n.* (尤指金、银的)铸块，锭

blister steel /ˈblɪstə(r) stiːl/ [冶]泡钢，泡面钢

cementation /ˌsiːmenˈteɪʃn/ *n.* (金属的)渗镀

charcoal / ˈtʃɑːkəʊl/ *n.* 炭，木炭(可作燃料或供作画)
crucible / ˈkruːsɪbl/ *n.* 熔炉
manganese / ˈmæŋgəniːz/ *n.* 锰
spiegeleisen / ˈspiːg(ə)lˌaɪz(ə)n/ *n.* 镜铁
phosphorus / ˈfɒsfərəs/ *n.* 磷
limestone / ˈlaɪmstəʊn/ *n.* 石灰岩
open-hearth / əʊpənˈhɑːθ/ *adj.* 使用平炉(炼钢)的

◆ **Notes**

1. **Sidney Gilchrist Thomas**：British metallurgist and inventor who discovered (1875) a method for eliminating phosphorus (a major impurity in some iron ores) in the Bessemer converter. The method is now called the Thomas-Gilchrist process, the Thomas process, or the basic process.

2. **The open-hearth process**：a process for making steel using an open-hearth furnace 平炉炼钢法

☞ **Section D　Language Focus**

Task One │ Text organization
Work in groups and discuss the organization of the text and fill in the blanks.

Parts	Paragraphs	Main Ideas
Part 1	Para. 1-3	Introduction to the 1._____ of steel
Part 2	Para. 4-8	Introduction to the era of 2._____
Part 3	Para. 9-13	The Bessemer process and modern 3._____
Part 4	Para. 14-16	The 4._____ process and the birth of the 5._____

Task Two │ Answer the following questions based on the information contained in the text.
1. The word "versatile" in the second paragraph means _____ .
　　A. competent in many areas
　　B. able to turn with ease from one thing to another
　　C. having various uses
　　D. changing rapidly

Unit 1 Introduction

2. What is the most breakthrough in steel history? _____
   A. Henry Bessemer found a solution to inefficient production processes.
   B. Henry Bessemer found a solution to iron's brittleness.
   C. Henry Bessemer developed an effective way to reduce oxygen to increase the production.
   D. Henry Bessemer used oxygen to reduce the carbon content in iron.

3. Blister steel production advanced in the 1740s, _____.
   A. when Benjamin Huntsman found high-quality steel for his clock springs
   B. when Benjamin Huntsman found that the metal could be melted easily
   C. when Benjamin Huntsman found that the metal could be refined in clay crucibles
   D. when Benjamin Huntsman found that the metal could be melted and refined to remove slag that the cementation process left behind

4. What did German engineer Karl Wilhelm Siemens do to increase steel production? _____
   A. By employing the Bessemer process.
   B. By making steel using an open-hearth furnace.
   C. By decreasing the cost of steel production.
   D. By frequent testing of the molten steel.

5. What is the author's purpose for writing this passage? Which of the following is Not correct? _____
   A. To provide readers with the history of steel.
   B. To introduce the Bessemer process.
   C. To teach readers to make steel.
   D. To show readers the open-hearth process.

Task Three | Fill in the blanks with the words or phrases given in the box. You may not use any of the words or phrases more than once.

The majority of global steel production, about 66%, is now produced in basic oxygen facilities — the development of a method to 1. _____ oxygen from nitrogen on an industrial scale in the 1960s allowed for major 2. _____ in the development of 3. _____ oxygen furnaces.

Basic oxygen furnaces blow oxygen into large 4. _____ of molten iron and scrap steel and can complete a charge much more quickly than open-hearth 5. _____. Large vessels 6. _____ up to 350 metric tons of iron can complete 7. _____ to steel in less than one hour.

The 8. _____ efficiencies of oxygen steelmaking made open-hearth factories uncompetitive and, following the 9. _____ of oxygen steelmaking in the 1960s, open-hearth operations began closing. The last open-hearth 10. _____ in the US closed in 1992 and China in 2001.

A. conversion	I. holding
B. separate	J. convert
C. quantities	K. discovery
D. advances	L. basic
E. conversation	M. methods
F. cost	N. facility
G. quality	O. start
H. advent	

Task Four | The following passage has three paragraphs. Choose the correct clause for each paragraph from the list of clauses below.

A. has been steadily increasing for over 50 years
B. were being used for the manufacturing of steel alloys
C. more than sufficient to heat steel production
D. get a pure form of steel from the scrap
E. occurred that would have a strong influence on the evolution of steel production

Just after the turn of the century, another development 1._____. Paul Heroult's electric arc furnace (EAF) was designed to pass an electric current through charged material, resulting in exothermic oxidation and temperatures up to 3272°F (1800°C), 2._____.

Initially used for specialty steels, EAFs grew in use and, by World War II, 3._____. The low investment cost involved in setting up EAF mills allowed them to compete with the major US producers like US Steel Corp. and Bethlehem Steel, especially in carbon steels, or long products.

Because EAFs can produce steel from 100% scrap, or cold ferrous, feed, less energy per unit of production is needed. As opposed to basic oxygen hearths, operations can also be stopped and started with a little-associated cost. For these reasons, production via EAFs 4._____ and now accounts for about 33% of global steel production.

Unit 1　Introduction

Task Five | Translate the following sentences into English, using the words or phrases in the brackets.

1. 钢的广泛应用证明了这种材料的多功能性。(versatility)

2. 油价上涨让人们更愿意使用公共交通工具。(incentive)

3. 两个商店老板在市中心的步行街购物中心租用了毗邻的店铺。(adjoining)

4. 实体零售店的关闭给零售业销售造成了不良影响。(deleterious)

5. 水是氢和氧的化合物。(compound)

Part Ⅱ: Text B

Section A Warm-up Questions

Discuss with your partner about the following questions.
1. What is the major impact of iron on human civilization?
2. Iron can be used for a wide variety of purposes, can you list some of them?
3. What are the properties of metals?

Section B Listening Practice

Listen to the passage and fill in the blanks below.

The closure of a steelmaking plant annoys the new president. Among the first visitors to the Elysee Palace after Franois Hollande's victory in the French presidential 1._____ last month were trade-union leaders from Arcelor Mittal's steel factory at Florange, in Lorraine.

The region is the crucible of France's 2._____ heavy industry: people have been making iron in Lorraine for more than 300 years. But the blast-furnace fires started to go out last year when 3._____ slashed demand for steel and forced Arcelor Mittal, the world's biggest steelmaker by some distance, to start closing its least 4._____ plants. By October both Florange's furnaces had been snuffed out, supposedly for a year. Last month the company said they would stay cool for another six months.

The unions fear that they will never be 5._____ up again and that steelmaking on a symbolic site will come to an end. After the union leaders' visit Mr Hollande asked Arnaud de Montebourg, the minister for 6._____ , to get an expert to review the prospects for Florange and to 7._____ the governments of Belgium, Germany, Luxembourg and Spain to "act together vis-a-vis ArcelorMittal"—whatever that means.

Union fears are probably well founded: Arcelor Mittal has 8._____ Florange and its plants at Liege, in Belgium, "non-core". The company has already started the cumbersome, two-year process needed under Belgian labour law to close Liege 9._____ . Although it describes the Florange shutdown as 10._____ , it plainly wants to close that plant for good too, although it is loth to say so in public.

Section C Active Reading

The History of Metallurgy

1 The history of metallurgy developed in an order related to the properties of the metals available to human beings, and to the ever increasing human knowledge of those properties, and of how to create environments of greater and greater temperature, enabling metals to be smelted, melted and alloyed with other metals. Metallurgical history began with the use of native metals, which are metals not attached to an ore. Such native metals are fairly rare so the widespread use of metals really began when humans learned how to **extract** metals from their ores, a process known as smelting.

2 It was not until the 14th century that iron smelting furnaces capable of melting iron were built in Europe. These furnaces were known as blast furnaces and were **substantially** larger than earlier furnaces. The blast furnaces had water **powered** bellows which produced much higher furnace temperatures as the bellows produced a continuous and strong flow of air through the **tuyeres** into the furnace. The higher temperatures allowed the iron to absorb a small quantity of carbon, which lowered the melting point of the iron to a temperature which the blast furnace could obtain. The melted iron, known as pig iron, could be poured into moulds or could be remelted and cast into any shape. The carbon in the pig iron could be removed to produce wrought iron which was more **malleable** than pig iron.

3 Substantial improvements were made to blast furnaces between 1500 and 1700. **Reverberatory furnaces**, with no chimneys and using underground pipes to bring in air achieved higher temperatures with **domed** roofs lined with **clay** reflecting the heat back into the furnace. Continuous smelting processes, which involved ore and fuel being continuously fed into the furnace to provide a continuous supply of iron greatly increased efficiency and production.

4 The use of coke, purified **bituminous** coal, in blast furnaces began around 1709 and greatly increased after 1760 when a method was found to get rid of **silicon** from iron produced from blast furnaces using coke. The silicon made it costly to convert pig iron into wrought iron. In the late 18th century coke replaced charcoal in most British blast furnaces. Blast furnaces produce pig iron but for many products the more malleable wrought iron was more suitable. The conversion of pig iron into wrought iron involved eliminating the carbon from the pig iron. An improved method of getting the carbon out of the pig iron was invented by Henry Cort in 1784. Cort's **puddling process** melted the pig iron in a reverberatory furnace which burnt the carbon and other **impurities** out of the iron and produced a mixture of iron and slag. The slag was removed by **hammering** to produce the wrought iron.

5 A further improvement to blast furnaces allowing still higher temperatures and reduced fuel use was invented by James Neilsen in 1829. Neilsen's invention involved using the furnaces own gases to **preheat** air before it entered the furnace. The air entered the furnace through a red-hot tube heated by the furnaces own gases and the hot air allowed the furnace to reach temperatures not previously obtainable. The preheating of the air blast was further improved by Edward Cowper in 1860 when he invented the hot-blast stove. Waste gases from the furnace were fed into a brick lined stove and heated the stove. Air entering the furnace is passed through the stove so it is heated before it reached the furnace.

6 Wrought iron was the principal material of the Industrial Revolution. Steel was a better material but was too expensive for widespread use during the Industrial Revolution. Steel is chemically midway between wrought iron, which contains almost no carbon, and pig iron which contains about 4% carbon. Steel usually contains between 0.2% carbon and 1.5% carbon. It was not until the second half of the 19th century that a process for creating cheap steel was invented. The **Bessemer process** was **patented** in 1856 and used a vessel called a **converter** into which molten pig iron was poured. Air was blown through holes in the base of the converter. The oxygen from the air combines with some of the iron to produce iron oxide which reacts with the carbon in the pig iron to produce carbon monoxide which releases some of the carbon from the pig iron. The remaining carbon is removed when the oxygen in the air is combined with silicon and manganese which form a **slag**. The resulting metal was brittle so manganese is added to remove the brittleness and then carbon is added to bring the steel up to the desired carbon content. The same process was independently invented in America by William Kelly.

7 An **alternative** method of making steel, known as the open-hearth process was invented in 1864 by William and Frederick Siemens and then improved by Pierre and Emile Martin. The open-hearth process involved pre-heating the air going into the furnace in two **chambers** that operated alternatively. The chambers, known as **regenerators**, contained a fire brick checker work and were alternatively heated by the furnace gases so the air passing into the furnace through the regenerators was heated resulting in higher furnace temperatures. As with the Bessemer process, iron oxide was used to remove carbon and other impurities, and manganese was added to remove brittleness and if necessary carbon was added to obtain the desired carbon levels.

8 The invention of electrical generators led to the use of electricity for heating furnaces. The first electric arc furnace began operation in 1902 and, while more expensive than the Bessemer and the open-hearth processes, was able to produce better quality steel due to it having fewer impurities than steel which had been in contact with fuel. Electric furnaces were able to produce greater heat and the temperatures could be more easily controlled than with ordinary furnaces. The use of electric

furnaces was to result in the large scale production of metals such as **tungsten**, **chromium** and manganese which when added to steel gave it useful properties such as improved hardness and resistance to wear. The electric furnace also allowed the mass production of aluminum. Aluminum is widespread on the Earth but it was difficult and expensive to extract from its ore, **bauxite**, before the invention of the electric furnace. The electric furnace produces aluminum by a process of high temperature electrolysis which produces molten aluminum in large quantities, although the process uses substantial quantities of electricity.

9 It had been long recognized that the use of oxygen, rather than air, in steel making would produce higher temperatures, faster production and reduce fuel costs. The high cost of producing oxygen stopped its use in steel making, until the price fell substantially and in 1948 the L-D process for using oxygen in steel making was developed. The L-D process involves blowing a jet of nearly pure oxygen at **supersonic** speed on to the surface of molten iron. The oxygen quickly burns out the carbon and other impurities resulting in faster production and reduced fuel costs.

10 The social and cultural consequences of the discovery of metallurgy were initially quite minor. Copper was initially used mainly for **ornaments** and jewelry as it was too soft a material to replace the stone tools and weapons used in **Neolithic** times. It was only when bronze was invented that metal tools and weapons replaced stone tools and weapons to create a Bronze Age. Bronze however was a reasonably expensive metal and when iron smelting was discovered the new metal soon replaced bronze as the principal material for tools and weapons. Iron ores are reasonably widespread and iron is a harder material than bronze, making it better for both tools and weapons.

✧ Words and Expressions

extract / ɪkˈstrækt/ v. to remove or obtain a substance from something, for example by using an industrial or a chemical process 提取；提炼

substantially / səbˈstænʃəli/ adv. very much; a lot 非常；大大地

powered / ˈpaʊəd/ adj. (usually in compounds) operated by a form of energy such as electricity or by the type of energy mentioned 由……驱动的；电动的

malleable / ˈmæliəbl/ adj. (of metal, etc.) that can be hit or pressed into different shapes easily without breaking or cracking 可锻造的；易成型的

domed / dəʊmd/ adj. having or shaped like a dome 圆顶状的；半球形的

clay / kleɪ/ n. a type of heavy, sticky earth that becomes hard when it is baked and is used to make things such as pots and bricks 黏土；陶土

impurity / ɪmˈpjʊərəti/ n. a substance that is present in small amounts in another substance, making it dirty or of poor quality 杂质

hammer / ˈhæmə(r) / *v.* to hit something with a hammer（用锤子）敲，锤打

preheat / ˌpriːˈhiːt / *v.* to heat（something, especially an oven or grill）to a particular temperature beforehand 预热

patent / ˈpætnt / *v.* to obtain a patent for an invention or a process 获得专利权

alternative / ɔːlˈtɜːnətɪv / *adj.* that can be used instead of something else 可供替代的

chamber / ˈtʃeɪmbə(r) / *n.* a room used for the particular purpose that is mentioned（作特殊用途的）房间，室

supersonic / ˌsuːpəˈsɒnɪk / *adj.* faster than the speed of sound 超声速的

ornament / ˈɔːnəmənt / *n.* (formal) an object that is worn as jewellery 首饰；饰物

Neolithic / ˌniːəˈlɪθɪk / *adj.* of the later part of the Stone Age 新石器时代的

◆ Terminology

tuyere / twiːˈjɛər / *n.* 鼓风口；风管嘴

reverberatory furnace 反射炉；反照炉

bituminous / bɪˈtjuːmɪnəs / *adj.* 含沥青的；沥青的

silicon / ˈsɪlɪkən / *n.* 硅

puddling process 搅炼法

Bessemer process（旧）酸性转炉炼钢法

converter / kənˈvɜːtə(r) / *n.* 转炉

slag / slæg / *n.* 矿渣；熔渣；炉渣

regenerator / rɪˈdʒenəreɪtə(r) / *n.* 再生器

tungsten / ˈtʌŋstən / *n.* 钨

chromium / ˈkrəʊmiəm / *n.* 铬

bauxite / ˈbɔːksaɪt / *n.* 铝土矿

☞ Section D Language Focus

Task One | Are the following statements True or False according to the passage? Write T/F accordingly.

1. Great improvements are being made to blast furnaces between 1500 and 1700.　　（　　）
2. Steel was a better material and inexpensive during the Industrial Revolution.　　（　　）
3. The open-hearth process is an alternative method of making steel.　　（　　）

Unit 1　Introduction

4. The use of air in making steel will result in higher temperatures, faster production and lower fuel costs.　　　　　　　　　　　　　　　　　　　　　　　　　(　　)

5. Copper was originally used mainly for decoration and jewelry because it was too soft to replace the stone tools and weapons.　　　　　　　　　　　　　　　(　　)

Task Two | Translate the following sentences into Chinese.

1. The higher temperatures allowed the iron to absorb a small quantity of carbon, which lowered the melting point of the iron to a temperature which the blast furnace could obtain.

2. Continuous smelting processes, which involved ore and fuel being continuously fed into the furnace to provide a continuous supply of iron greatly increased efficiency and production.

3. The slag was removed by hammering to produce the wrought iron.

4. The invention of electrical generators led to the use of electricity for heating furnaces.

5. It was only when bronze was invented that metal tools and weapons replaced stone tools and weapons to create a Bronze Age.

Task Three | Choose the best answers for each blank of the following passage.

The ultimate cause of much historical, social and cultural change is the gradual accumulation of human knowledge of the environment. Human beings use the materials in their environment, including fire and metals, to meet their needs and 1. _____ human knowledge of fire and metals enables human needs to be met in a more efficient manner.

Fire and metals have particular 2. _____ and human knowledge of those properties increases over time in a particular order. Increasing human knowledge of how to create higher and higher temperatures enables the 3. _____ and melting of a wider range of ores and metals. Those ores and metals that could be smelted and 4. _____ at lower temperatures were used before the ores and metals which had higher smelting and melting points. This 5. _____ that copper, and its alloy bronze, were used before iron and its alloy steel. Pure metals, like copper and iron, were used before alloys such as, bronze and steel, as the 6. _____ of alloys is more complicated than the manufacture of pure metals.

The simplest knowledge is 7. _____ first and more complex knowledge is acquired later. The order of discovery determines the 8. _____ of human social and cultural history, as knowledge of new and more efficient means of smelting ores and melting metals, results in new technology, which 9. _____ to the development of new social and ideological systems. This means human social and cultural history, had to follow a 10. _____ course, a course that was determined by the properties of the materials in the human environment.

1. A. increased   B. increasingly   C. decreased   D. decreasing
2. A. possibilities B. properly   C. properties   D. proper
3. A. smelt   B. smelting   C. smell   D. smells
4. A. melt   B. melted   C. heat   D. heating
5. A. meant   B. means   C. advises   D. suggests
6. A. structure   B. fabricate   C. manufacture   D. fracture
7. A. acquired   B. required   C. inquired   D. questioned
8. A. course   B. method   C. research   D. quest
9. A. distributes   B. distribution   C. contribution   D. contributes
10. A. deliberate   B. purposeful   C. general   D. particular

Unit 1 Introduction

Task Four | The following passage has three paragraphs. Choose the correct sentence for each paragraph from the list of sentences below.

A. If metals did not exist at all then we would be restricted to stone, bone and wood tools.
B. It is hard to conceive of wooden or stone steam engines or internal combustion engines.
C. The use of iron tools and weapons gave humankind greater control of their environment leading to increased population and larger settlements.
D. If the smelting and melting points of metals were different then human history would have been different.
E. Cheap steel replaced iron in a great variety of applications.

Iron was used for a wide variety of purposes such as nails and tools, cooking pots and kitchen utensils. 1._____ Iron became the principal material for the Industrial Revolution being used in steam engines, industrial machinery, in railways for rails and locomotives, for bridges, buildings and in iron ships.

The Bessemer and open-hearth steel making processes led to a great reduction in the price and increase in production of steel. 2._____ Steel was used in railways and for ships and in bridge building. Motor vehicles became one of the biggest users of steel in the 20th century and different types of steel began to be developed for different purposes. Stainless steel containing chromium and nickel was widely used in kitchens and in industrial plants vulnerable to corrosion as it does not rust.

Metallurgy has had a great effect on human societies, certainly since the Bronze Age. 3._____ This would have had an enormous effect on human history. It is doubtful whether the Industrial Revolution and the industrial world that emerged from it, would have been possible without metals. 4._____ Wooden engines would catch fire while it is doubtful that stone could be worked in a way that could create pistons and cylinders. Without metals it is doubtful that there would be usable electricity, as the transfer of electricity over significant distances would be difficult or impossible.

Part Ⅲ: Academic Skills

Academic Writing Skills(Ⅰ) *

1. What is Academic Writing?

Academic writing differs from other types of writing such as journalistic or creative writing. In most forms of academic writing a detached and objective approach is required. An academic argument appeals to logic and provides evidence in support of an intellectual position. It is important to present your arguments in logical order and to arrive at conclusions. However, academic writing can take many forms. You may be asked to write an essay, a report, a review or a reflective article. Different styles adhere to each of these types of academic writing, so always check with your lecturer. In academic writing, writers always interact with each others' texts and so there will be frequent references to the ideas, thinking or research of other authors' writing in this field.

You must give credit to those with whom you are interacting and there are structured guidelines for referencing and citation. Also, in academic writing it is important that when a claim is made it is backed up by reasons based on some form of evidence; it is expected that the author takes a critical approach to the material being explored.

2. Planning for Your Writing Task

Writing typically consists of 4 main stages: planning, writing, editing and reviewing. As writing is an iterative process, these activities do not occur in a fixed order; rather, writers move among these activities although typically, more time is spent on planning or thinking at the start and on editing and reviewing at the end. Planning for your writing has been identified as one of the key practices of good writers and you need to factor in time to gather, absorb and plan your arguments before composing text.

1) Freewriting

Freewriting involves writing in full sentences about a topic for a specified amount of time without planning or worrying about quality; it can help writing at all stages. Elbow & Belanof (2003) argue that freewriting is about trusting yourself and your words; they believe it is especially helpful at the

---

* Taken from *Developing Your Academic Writing Skills: A Handbook* by Marian Fitzmaurice & Ciara O'Farrell.

Unit 1 Introduction

initial stages of academic writing.

Freewriting means:
- Not showing your words to anyone (unless you later change your mind);
- Not having to stay on one topic—that is, freely digressing;
- Not thinking about spelling, grammar, and mechanics;
- Not worrying about how good the writing is.

Most people learn and practise freewriting by doing freewriting exercises of five to ten minutes. It is more than just putting words on paper as it helps improve thinking and also this is the beginning of your voice in the writing. It is really good to do some freewriting or focused freewriting which requires you to stay on one topic on a regular basis. So try it and remember the important thing is to keep writing.

2) Using primary evidence or published sources

When reading and gathering information in an academic context, evidence comes from two sources, primary and published, although for much undergraduate writing it is acceptable to use published sources only. Primary evidence is the raw data such as questionnaires, interviews, focus groups or experiments that are used by a researcher to gather data to answer a specific research question; they provide proof or insight in regard to the topic or question.

Published sources constitute the literature on a topic, such as books, journals or reports. In journals, published sources from peer-reviewed journals carry most weight. To be published in a peer-reviewed journal, the paper is typically sent out to two or three experts in the field for review and is only published when the reviewers and editor deem it suitable. It is important to read appropriate peer reviewed journals in your literature when planning your academic writing. Ask your lecturer to recommend some. Further, each academic subject has a Subject Librarian who is very willing to provide training in using the library catalogue and accessing resources and relevant databases. Make time to speak to the Subject Librarian who is there to help you.

Activity 1: Getting started—4 things to write.

1. First of all, consider what is your topic for consideration. Write this in less than 25 words.
2. Next, brainstorm all the ideas that come to mind. Let your ideas flow and write down everything. Don't censor.
3. Now, do some freewriting to a prompt: What do I already know about this topic/question? What do I need to find out? Write for 5 minutes.
4. Finally, write a list of books, journals, reports that you need to read. This will help you provide the foundation for your writing /assignment/project.

3) Referencing the work of others in one's own writing

In academic writing, you will almost always draw on the work of other writers: knowing how to reference is key. There are a variety of referencing systems but in all systems, a source is cited in the text with a name or number. The name or number connects with the full source details in a footnote or reference list.

4) Structure and sequence

It is useful to think about the structure of an assignment. Simple as it might seem, all written assignments should have three parts:

- Introduction
- Main Body
- Conclusion

Sections	Ideas and Arguments	Words
Introduction	State the main issues and the issues you will deal with in the paper	10% to 15%
Main Body	Readings	70% to 80%
Conclusion	This should not be a summary of all the points made in the assignment/paper but it should state what you believe to be reasonable conclusions based on the arguments made. It is important to relate the conclusions to the question.	10% to 15%

Completing your assignment: Your writing should contain a strong, coherent argument.

Activity 2

To help writers achieve an authoritative stance in their introduction, Thompson and Kamler (2013) suggest selecting a passage of writing from the introduction of a published article and deleting its content. What remains is the skeleton that writers can then work with. The example below shows how the writer builds a connection with the field and structures the article. Removing the content makes the syntax visible without plagiarising. "It makes explicit the ways of arguing and locating used in particular discourses". However, do not feel obliged to follow a particular structure if you have your own approach.

Unit 1   Introduction

In this article, I discuss the main arguments that deal with the issue of _____ _____.

In distinguishing between _____ it is my purpose to highlight _____, by pointing to _____ _____.

Besides providing a map of the _____, I assess the extent to which these _____ lay a ground work for _____.

The article is structured as follows. After giving an overview of the scope of the _____ _____, I review the particular _____ _____.

Next, I provide a summary of _____.

Finally, in the last two sections, I consider several implications derived from _____ _____ and argue that _____.

Part Ⅳ: Extended Reading and Translation

Translate the following passage into Chinese.

China's steel firms registered a third consecutive monthly increase in exports in November, driven by rebounding demand in overseas markets, according to the Ministry of Industry and Information Technology.

The country's steel exports rose 9 percent month-on-month to stand at 4.4 million tons in November and are expected to further increase in December, the ministry said.

The competitive advantages of China's steel prices in the international market and the pick-up of overseas demand contributed to the increase in China's steel exports, according to the ministry.

China exported about 48.83 million tons of steel in the first 11 months, down 18.1 percent year-on-year, while the country's imports jumped 74.3 percent year-on-year to 18.86 million tons during the same period. The country's iron ore imports rose 10.9 percent year-on-year to 1.07 billion tons in the January-November period, exceeding the amount imported in the whole of 2019, the Ministry added.

Unit 2 Iron Making

Part Ⅰ: Text A

Section A Warm-up Questions

Discuss with your partner about the following questions.
1. How much do you know about the metal fabrication methods?
2. What is forging?
3. Which is the strongest material in the world?

Section B Listening Practice

Listen to the passage and fill in the blanks below.

Researchers say they have created a new super-material in the laboratory. They say it is stronger, lighter and 1._____. But the new material is not a product of high technology or a metallic substance. It is wood. Researchers Liangbing Hu and Teng Li are with the University of Maryland in the United States.

By 2._____, heat and pressure, they have made wood three times denser and 10 times stronger. Hu says that means it can 3._____ some of the world's strongest materials. "We're interested in replacing steel and carbon fibers with strong wood structures," he said.

The process does not require any special raw, 4._____. Hu said that means the cost of the product will be relatively low. He added, "We can start with very cheap wood, and we can also start with 5._____. But in the end, ultimately we get very similar performance."

The researchers are calling their product "super wood". 6._____ removing a natural polymer called lignin. Hu explained that "lignin is like a binder to hold all the components together in natural wood. In our process, we found out 7._____ the wood completely, we have to remove some of these binders."

Strong chemicals are put on the wood. The chemicals remove about half of its lignin. Then, 8._____ and pressurized for a day, the treated wood is ready. The researchers say the wood is strong enough to 9._____, airplanes, wind turbines and more. A report on their findings was published in the journal *Nature*. The researchers say super-strength wood would have 10._____ than the steel or other metals it could replace. However, their process is not pollution-free. It uses some of the same chemicals involved in making paper. Researcher Teng Li says he and Hu are finding many new ways to use wood, including creating batteries and see-through wood and paper.

Section C   Active Reading

Types of Metal Fabrication Processes

1   **Metal fabrication** is a broad term referring to any process that cuts, shapes, or molds metal material into a final product. Instead of an end product being **assembled** from ready-made components, fabrication creates an end product from raw or semi-finished materials. Choosing a fabrication method suited to a given project depends on the product's intended purpose and the materials used in **crafting** it. Common custom metal fabrication processes are as follows:

2   Casting is when molten metal is poured a mold or **die** and allowed to cool and harden into the desired shape. The process is ideal for mass-production of parts with the reuse of the same mold to create identical products. There are several different types of casting. Die-casting is when liquid metal is forced into a die instead of a mold, and there the applied pressure keeps it in place until it hardens. This process is known for the high-speed applications it supports. Permanent mold casting involves pouring the molten metal into a mold.

3   This very common type of fabrication is the cutting of a metal **workpiece** to split it into smaller sections. While **sawing** is the oldest method of cutting, modern methods include laser cutting, **waterjet** cutting, power scissors, and **plasma arc cutting**. There are many different methods of cutting, from manual and power tools to **Computer Numerical Control** (**CNC**) cutters. Cutting may be the first stage in a longer fabrication process or the only process used.

Unit 2 Iron Making

4 Drawing uses **tensile** force to pull metal into and through a **tapered** die. The die stretches the metal into a thinner shape. Usually drawing is performed at room temperature, and is called cold drawing, but the metal workpiece can be heated in order to reduce the required force. This process is considered deep drawing when the end product has a depth that is equal to or greater than its **radius**. It is usually used with sheet metal fabrication to turn sheets of metal into **hollow cylindrical** or box-shaped **vessels**.

5 Metal is manipulated to bend at an angle. The most common means is with a **brake** press that creates **creases** in the metal by **pinching** it. The workpiece is held between a punch and a die and forced to crease with pressure from the punch. This process is usually used to shape sheet metal. **Folding** can also be done by hammering the metal until it bends, or by using a folding machine, also known as a folder. The machine has a flat surface where the sheet metal is placed, a clamping bar that holds the workpiece in place, and a front panel that lifts upwards and forces the metal extended over it to bend.

6 Forging uses **compressive** force to shape metal. A hammer or die strikes the metal workpiece until the desired shape is formed. This process can be done with the metal at room temperature and is called cold forging. Forging can also be performed with the metal heated to a range of above room temperature to below the **recrystallization** temperature and is then called warm forging. When the metal is heated to its recrystallization temperature, which varies by metal, the process is called hot forging. Forging is one of the oldest types of fabrication, with blacksmiths using forging centuries ago.

7 In the extrusion process, the workpiece is forced through or around an open or closed die. When forced through an open or closed die, the **diameter** of the workpiece is reduced to the cross-section of the die. When pressed around a die, a **cavity** is formed within the workpiece. Both of these processes generally use a metal slug or cylinder as the workpiece, and a **ram** to perform the impact operation. The resulting cylindrical item product is often wiring or piping. The die cross-section can have different shapes to produce differently shaped parts. Extrusion can be continuous to create very long pieces, or semi-continuous in order to create many shorter pieces.

8 Also called cold **extrusion**, impact extrusion is performed at room temperature and increases the strength of the part, making it stronger than the original material. When enough force is applied to the appropriate metal, it starts to flow into the available shape, much like the movement of viscous liquid. Cold extrusion is commonly used for steel fabrication. Hot extrusion is performed at an increased temperature, to keep the metal from **hardening** and to make it easier to push through the die. It's usually used for copper fabrication, as well to create custom aluminum parts.

9 Machining refers to the process of shaping metal by removing the unwanted material from it. This

process can be performed in a variety of ways. There are many different machining processes, including **drilling**, turning, and **milling**.

10 Uniquely shaped turrets on a punch press hit the metal through or into a die to create holes. The end product can either be the piece of metal with holes for fastening purposes, or it can be the now removed, and shaped metal pieces called the blanking. Most punch presses are mechanical but smaller and simpler punches can be hand-powered. CNC punch presses are also now common and are used for both heavy and light metal work.

11 One long, straight cut is achieved by combining two tools, with one of the tools above the metal and other one located below for applying pressure. The upper blade forces the metal down onto the stationary lower blade and **fractures** it. The fracture then spreads inward for complete separation. The sheared edges are usually **burred.** It is ideal for cutting smaller lengths and differently shaped materials since the blades can be mounted at angles to reduce the required force.

12 This process is similar to **punching**, except the press doesn't create a hole in the metal, but an **indentation**. The turret doesn't completely force the metal through the die, but only raises it. **Stamping** is used to form shapes, letters, or images in a metal panel or sheet. Mechanical and **hydraulic** are the two types of stamping presses. Metal stamping machines cast, punch, cut and shape metal sheets. Sheets of up to 1/4 inch thickness are molded into specified shapes and sizes. The presses used for metal stampings can create a wide range of products, and they can perform a series of operations including blanking, metal coining, and four slide forming. Metal coining (as the title implies) can be used to create coins, but it has other uses as well, such as parts for electronics. Four slide forming incorporates a variety of stamping and forming processes to create more complex products, and it is particularly effective for smaller parts.

13 With **welding**, two or more pieces of metal are joined together, through a combination of heat and pressure. This is a popular process because the pieces of metal can be any shape or size. Four of the popular types of welding procedures are Stick or Arc Welding, Metal Inert Gas Welding, Tungsten Inert Gas Welding, and Flux Cored Arc Welding.

◆ Words and Expressions

fabrication / ˌfæbrɪˈkeɪʃn/ n. the act of making or producing goods, equipment, etc. from various different materials 制造；装配；组装

assemble / əˈsembl/ v. to fit together all the separate parts of something, for example a piece of furniture 装配；组装

craft / krɑːft/ v. to make something using special skills, especially with your hands(尤指用

Unit 2 Iron Making

手工)精心制作

saw / sɔː/ *n.* a tool for cutting wood, which has a blade with sharp teeth along one edge 锯

tensile / ˈtensaɪl/ *adj.* used to describe the extent to which something can stretch without breaking 张力的;拉力的;抗张的

radius / ˈreɪdiəs/ *n.* a straight line between the centre of a circle and any point on its outer edge; the length of this line 半径(长度)

hollow / ˈhɒləʊ/ *adj.* having a hole or empty space inside 中空的;空心的

vessel /ˈvesl/ *n.* a container used for holding liquids, such as a bowl, cup, etc. 容器

brake /breɪk/ *n.* a device for slowing or stopping a vehicle 制动器;车闸

crease / kriːs/ *n.* an untidy line that is made in cloth or paper when it is pressed or crushed 褶痕;皱痕

pinch / pɪntʃ/ *v.* to hold something tightly between the thumb and finger or between two things that are pressed together 捏住;夹紧

fold / fəʊld/ *v.* to bend something so that it becomes smaller or flatter and can be stored or carried more easily; to bend or be able to bend in this way 折小,叠平

compressive / kəmˈpresɪv/ *adj.* compressing or having the power or capacity to compress 压缩的;有压缩力的

diameter / daɪˈæmɪtə(r)/ *n.* The length of a straight line that can be drawn across a round object, passing through the middle of it 直径

cavity / ˈkævəti/ *n.* a hole or empty space inside something solid 洞;孔

extrusion / ɪkˈstruːʒn/ *n.* the act or process of extruding something 挤压

harden / ˈhɑːdn/ *v.* to become or make something become firm, stiff or solid (使)变硬,硬化

drill / drɪl/ *v.* to make a hole in something, using a drill 钻(孔);打(眼)

mill / mɪl/ *v.* to crush or grind something in a mill (用磨粉机)碾碎,磨成粉

fracture / ˈfræktʃə(r)/ *v.* to break or crack; to make something break or crack (使)断裂,折断,破裂

burr / bɜː(r)/ *v.* the soft regular noise made by parts of a machine moving quickly (机器部件快速运转时有规律的)呼呼声

punch / pʌntʃ/ *v.* to make a hole in something with a punch or some other sharp object(用打孔器等)打孔

indentation / ˌɪndenˈteɪʃn/ *n.* a cut or mark on the edge or surface of something 凹陷

stamp / stæmp/ *v.* to cut and shape an object from a piece of metal or plastic using a special machine or tool 冲压

weld / weld/ *v.* to join pieces of metal together by heating their edges and pressing them together 焊接；熔接

◆ Terminology

die / daɪ/ *n.* 模具；冲模；压模
workpiece / ˈwɜːkpiːs/ *n.* 工作部件；工件
waterjet / ˈwɔːtədʒet/ *n.* 喷水
plasma arc cutting 等离子弧切割
Computer Numerical Control 电脑数控
tapered / ˈteɪpəd/ *adj.* 锥形的
cylindrical / səˈlɪndrɪkl/ *adj.* 圆柱形的；圆筒状的
recrystallization / riːˌkrɪstəlaɪˈzeɪʃən/ *n.* 再结晶
ram / ræm/ *n.* 撞击装置
hydraulic / haɪˈdrɒlɪk/ *adj.* 液压驱动的

☞ Section D Language Focus

Task One | Text organization

Work in groups and discuss the organization of the text and fill in the blanks.

Parts	Paragraphs	Main Ideas
Part 1	Para. 1	Introduction to 1. _____
Part 2	Para. 2–13	Introduction to 2. _____

Task Two | Answer the following questions based on the information contained in the text.

1. The word "recrystallization" in the sixth paragraph means _____.

    A. the degree of hotness

    B. the degree of coldness

    C. increase the strength

    D. a technique used to purify chemicals

2. What do you know about sand casting? _____

    A. It is faster than the other forms of casting.

Unit 2 Iron Making

  B. It is more costly than the other forms of casting.

  C. It is generally more economical than the other forms of casting.

  D. All of the above are right.

3. What is the method mentioned in the fifth paragraph? _____

  A. Forging.

  B. Pinching.

  C. Punching.

  D. Folding.

4. According to the author, which of the following types of fabrication is the oldest? _____

  A. Extrusion.

  B. Forging.

  C. Punching.

  D. Stamping.

5. What is the main idea of this text? _____

  A. To provide readers with the history of metal fabrication.

  B. To introduce metal fabricators.

  C. To introduce metal fabricated products.

  D. To give an overview of common fabrication methods.

Task Three | Fill in the blanks with the words or phrases given in the box. You may not use any of the words or phrases more than once.

Modern metallurgy is divided into two subtypes. Process metallurgy refers to the steps 1. _____ in producing metals, in most cases, from sulfides or oxides, and then refining them in their reduced form through electrolysis or selective oxidation of impurities. Physical metallurgy studies the structure of metals, 2. _____ on their composition and treatment, and how this structure is related to their 3. _____ . It is also 4. _____ with the scientific principles and engineering applications employed in metals fabrication and treatments, and how metal products hold up under their industrial usages.

Metallurgical engineers 5. _____ different forms of metal testing. In that way, they can make quantified 6. _____ about a metal's strength. These tests are meant to determine such properties as metal hardness, impact toughness, and tensile strength, to name of few.

In 7. _____ , elemental metals, in their pure native form, are too soft for industrial uses. That is why the science of metallurgy tends to 8. _____ on the manufacture of alloys, in which metals are 9. _____ together or with non-metals. Steel and cast irons are examples of iron-carbon 10._____. Aluminum, copper, iron, magnesium, and zinc are the metals that are used most, usually in their alloy forms.

A. based	I. involved
B. assumptions	J. employ
C. properties	K. removed
D. alloys	L. concerned
E. combination	M. generally
F. general	N. related
G. focus	O. combined
H. aims	

Task Four | The following passage has three paragraphs. Choose the correct clause for each paragraph from the list of clauses below.

A. CNC machining is generally used
B. This process can be done manually
C. The cutting tool can be angled differently
D. To increase the strength or to obtain special properties
E. Drilling uses a rotary cutting tool

1._____, the drill bit, to cut a hole in the material. The drill bit presses against the metal while being rotated very quickly in order to create a circular hole.

Turning uses a lathe to rotate the metal while a cutting tool moves in a linear motion to remove metal along the diameter, creating a cylindrical shape. 2._____ to create different shapes. It can be done manually or with a Computer Numerical Control (CNC) turning machine. 3._____ when part measurements must be extremely precise.

Milling uses rotating multi-point cutting tools to progressively remove material from the workpiece until the desired shape is achieved. The metal is slowly fed into the rotating cutting tool, or the tool is moved across the stationary metal, or both the workpiece and the tool are moved in relation to each other. 4._____ or with a CNC milling machine. Milling is often a secondary or finishing process, but it can be used as the sole method of fabrication from start to finish. The different types of milling include face milling, plain milling, angular milling, climb milling, and form milling.

Task Five | Translate the following sentences into English, using the words or phrases in the brackets.

1. 器件制造工艺简单。(fabrication)

2. 一些学生在业余时间喜欢装配模型火车。(assemble)

3. 宁有公开的敌人，不要虚伪的朋友。(hollow)

4. 您需要多花一些时间去加固您的系统以避免此类攻击。(harden)

5. 玛丽想在她家客厅的墙上钻个洞。(drill)

Part II: Text B

☞ Section A Warm-up Questions

Discuss with your partner about the following questions.
1. Do you know how iron was made?
2. Why is it important to develop iron industry?
3. How to reduce pollution while promoting iron industry?

☞ Section B Listening Practice

Listen to the passage and fill in the blanks below.

Building houses and offices out of toxic waste sounds like a pretty eccentric idea. Yet it may become 1. _____ if Ana Andres of the University of Cantabria in Spain has her way. Dr. Andres and her colleagues suggest, in Industrial & Engineering Chemistry Research, that the humble brick need not be made of pure clay. Instead, 2. _____ 30% of its weight could be slag—the toxic gunk left over when steel is made. Waelz slag, to give its technical name, is composed mainly of silica but is also 3. _____ in poisonous metals like lead and zinc. 4. _____ it safely is thus a problem. Getting rid of it usefully might sound like a miracle. But that is what Dr. Andres 5. _____.

A series of experiments she has 6. _____ over the past three years suggests this is not only possible but will make bricks cheaper and more 7. _____. Her research started after she read of previous work which had shown that many ceramics suffer 8. _____ when the clay used to make them is mixed with other materials, and that the molecular structure of some ceramics acts to trap atoms of toxic heavy metals.

She wondered whether these things might be true of brick clay and Waelz slag, and she began 9. _____. The answer, she found, was that they are. Bricks show no loss of useful mechanical properties even when 20%~30% of their content is slag. Nor do they leak. To check that, Dr. Andres and her team ground their bricks into powder and soaked them in water, shook them in special machines for days at a time, and even 10. _____ them in nitric acid. The pollutants stayed resolutely put.

Unit 2　Iron Making

Section C　Active Reading

How Iron Was Made

Introduction

1　Iron making evolved over a few thousand years. Using the ancient "**bloomery**" method, iron ore was converted directly into wrought iron by heating the ore while at the same time melting the ore's impurities and squeezing them out with hand hammers. This is also called the "direct process". By the 1100s water-powered hammers replaced hand hammers for **forging** out bars of iron.

2　In the late 1300s, some theorize that because of the **ravages** of the **plague** upon the labor forces in Europe, water power began to replace human or animal power to blow the air into iron making furnaces. Using water-powered **bellows**, a large and consistent volume of air created enough heat to completely melt the ore from which iron was made. The technology created two major developments in iron production. First, blast furnaces were now able to make cast iron for the production of **hollowware** such as pots and kettles. Second, in the new "indirect process" cast iron could be converted into wrought iron with a higher yield of iron from the ore than the direct process.

3　It is the indirect process that was brought to Massachusetts and spread through North America by the skilled iron makers that came to Saugus[2]. With improvements, the direct process continued as well and a few of the second generation iron making plants after Saugus thrived in the rural economy using the older, bloomery method.

Smelting

4　In 1646, the original blast furnace roared to life, lit with a 3000℃ fire that was kept burning 24 hours a day for months at a time. The blast furnace is where bog ore was smelted to create cast iron "pig" bars, so named because liquid cast iron was fed from a larger **trench** into smaller trenches as a mother sow to **suckling** pigs. To make cast iron, three raw materials were brought over the charging bridge and loaded into the chimney of the furnace.

5　Charcoal fueled a fire that burned hot enough to smelt the ore. Charcoal production was very labor intensive and required the work of many woodcutters, carters, and the **colliers** who **oversaw** the conversion of **seasoned** wood into charcoal.

6　**Bog ore** is an iron-rich **sedimentary** rock that was harvested locally from bogs and similar bodies of water. It was also found in fields and meadows that used to be bogs. Bog ore is often considerably less than 50% iron. The rest of the rock was made up of impurities that the workers had to remove. **Gabbro** was used as a flux; a way to purify the ore. It was mined on nearby Nahant and transported up the Saugus River by boat.

7   A wood fire was started at the bottom of the furnace to dry out the **mortar** that was between the new lining stones and brick. Gradually at first, charcoal, iron ore, and gabbro were charged into the top of the furnace in layers by the furnace fillers. The "burden" as it was called, was carefully managed by the founder. The founder also was responsible for managing air flow from the bellows.

8   The burden was held in place above the crucible (where the molten iron was collected) at the bottom of the furnace by a narrowing of the furnace lining called the "**boshes**". Air was pumped into the furnace above the crucible but below the boshes.

9   Air, though invisible, was also a raw material and blew into the furnace using large, water-powered bellows. Oxygen in the air made the fire burn hot and (managed properly) created the appropriate conditions for carbon monoxide to remove oxygen from the iron ore. As the air passed upward through the burden, it first came across the charcoal. As the charcoal burned, the air converted to carbon monoxide. Carbon monoxide continued to rise upward. It latched onto the oxygen atoms in the ore and was carried further upward and out the furnace stack as carbon dioxide.

10   High heat that was generated by the fire caused the gabbro to melt and form a flux. Flux performed multiple functions. Since it melted at a lower temperature than the iron in the ore, it facilitated the flow of **silicates** and other impurities from the ore. The **glassy** flux also coated the iron as it melted. It formed a protective barrier between the liquid iron and the oxygen in the furnace gasses and prevented the iron from **oxidizing** away.

11   As a charge was transformed, the flux with its impurities descended past the boshes into the crucible. The liquid iron that was coated in flux trickled past the boshes, through the slag, and settled to the bottom of the crucible. The displaced liquid slag floated on top of the molten iron along with any unburned charcoal bits, ash, and other dross.

12   Iron was graded as gray, white, or **mottled** and was checked by fracture testing, that is, breaking the iron to visually inspect the way that carbon was interspersed through the cast iron. The crystallization that produced the various grades was deliberately controlled by the founder. With great knowledge and skill, he regulated ore, fuel, air, flux, and even cooling rate to create desired attributes in the iron.

Casting

13   The casting shed at the base of the furnace is where the cast iron and slag waste were removed from the furnace. Molds were specially prepared and awaited the molten metal.

14   Gray iron was poured into molds composed of a clay/sand mix "**loam**" to make cast iron cookware like pots and kettles. Molds had to be carefully dried to reduce the risk of exploding steam pockets when the moist mold came in constant with molten iron. Gray iron was also cast in the sand

to make firebacks. When poured into the molds, it was necessary to separate the slag from the iron to keep the slag from being trapped in the iron. Cookware was finished by filing and cleaning before it was brought to the river for shipment.

15 Mottled iron was also cast into the sand in the shape of long bars. In this case the iron would be cast with slag and all and the slag would float to the top of the bars where it would break off. The iron then froze into heavy ingots or "pigs". Pig iron was an intermediate step in the making of wrought iron. Pig bars were dragged to the forge using oxen.

16 The slag waste **solidified** when it cooled and sometimes resembled glass. The slag was disposed of at the **waterfront** by dumping it over the bulkhead. Over time the slag pile grew. The slag pile remains today and when the archeologist was searching for the furnace, he traced the slag pile back to its origin at the furnace.

◆ Words and Expressions

forge / fɔːdʒ/ v. to shape metal by heating it in a fire and hitting it with a hammer; to make an object in this way 锻造；制作

ravage / 'rævɪdʒ/ n. a destructive action 毁坏；损坏

plague / pleɪɡ/ n. any infectious disease that kills a lot of people 瘟疫

hollowware / 'hɒləʊˌweə/ n. hollow articles made of metal, china, etc., such as pots, jugs, and kettles (由金属、瓷器等制成的)空心制品

trench / trentʃ/ n. a long deep hole dug in the ground, for example for carrying away water 沟；渠

suckling / 'sʌklɪŋ/ n. (old-fashioned) a baby or young animal that is still drinking milk from its mother 乳儿；乳兽

collier / 'kɒliə(r)/ n. a coal miner 煤矿工人

oversee / ˌəʊvə'siː/ v. to watch somebody/something and make sure that a job or an activity is done correctly 监督；监视

seasoned / 'siːznd/ adj. (of wood) made suitable for use by being left outside 风干的，晾干的(可加工使用)

sedimentary / ˌsedɪ'mentri/ adj. (geology) connected with or formed from the sand, stones, mud, etc. that settle at the bottom of lakes, etc. 沉积的；沉积形成的

glassy / 'ɡlɑːsi/ adj. like glass; smooth and shiny 光亮透明的

oxidize / 'ɒksɪdaɪz/ v. (chemistry) to remove one or more electrons from a substance, or to combine or to make something combine with oxygen, especially when this causes metal to become covered with rust(使)氧化;(尤指使)生锈

mottled / ˈmɒtld/ *adj.* marked with shapes of different colors without a regular pattern 斑驳的;杂色的

solidify / səˈlɪdɪfaɪ/ *v.* to become solid; to make sth. solid(使)凝固,变结实

waterfront / ˈwɔːtəfrʌnt/ *n.* a part of a town or an area that is next to water, for example in a harbour 滨水区;码头区

◆ Terminology

bloomery / ˈbluːməri/ *n.* 锻铁炉;熟铁块吹炼法

bellows / ˈbeləʊz/ *n.* 风箱;吹风器

bog ore / bɒg ɔː(r) / *n.* 沼矿

gabbro / ˈɡæbrəʊ/ *n.* 辉长岩

mortar / ˈmɔːtə(r)/ *n.* 灰泥;砂浆

bosh / bɒʃ/ *n.* (高炉风嘴以上的)炉腹

silicate /ˈsɪlɪkeɪt/ *n.* 硅酸盐

loam / ləʊm/ *n.* 壤土;肥土

◆ Notes

1 **Curtis White**: Former Supervisory Park Ranger at Saugus Iron Works National Historic Site

2 **Saugus**: town, northeastern Massachusetts. It was settled in 1629. The Saugus Iron Works (1646; now a national historic site) was the first successful ironworks and steelworks in colonial America. Saugus is now primarily residential with services, retail trade, some light manufacturing, and lobster fishing. 索格斯

3 **Nahant**: The Town of Nahant is a resort town of rocky coasts in the southernmost part of Essex County. Located in Eastern Massachusetts, located on a peninsula jutting south of Lynn and surrounded on three sides by the Atlantic Ocean. 纳汉特

Unit 2　Iron Making

👉 Section D　Language Focus

Task One | Are the following statements True or False according to the passage? Write T/F accordingly.

1. By the 1100s water power began to replace human or animal power to blow the air into iron making furnaces.　　　　　　　　　　　　　　　　　　　　　　　　(　)

2. It is the direct process that was brought to Massachusetts and spread through North America.
　　　　　　　　　　　　　　　　　　　　　　　　　　　　　　　　(　)

3. Air was a raw material and blew into the furnace using large, water-powered bellows.
　　　　　　　　　　　　　　　　　　　　　　　　　　　　　　　　(　)

4. The slag was graded as gray, white, and mottled.　　　　　　　　(　)

5. Pig iron is an intermediate process in the production of wrought iron.　(　)

Task Two | Translate the following sentences into Chinese.

1. By the 1100s water-powered hammers replaced hand hammers for forging out bars of iron.

2. Charcoal fueled a fire that burned hot enough to smelt the ore.

3. As the charcoal burned, the air converted to carbon monoxide.

4. With great knowledge and skill, he regulated ore, fuel, air, flux, and even cooling rate to create desired attributes in the iron.

5. Pig iron was an intermediate step in the making of wrought iron.

Task Three | Choose the best answers for each blank of the following passage.

Details regarding the original design of finery hearths have yet to be discovered. 1._____, they were specially constructed with stone and lined with cast iron plates. It is 2._____ that gray cast irons and white cast irons were both 3._____ by positioning the iron plates and 4._____ the air draft from water-powered bellows. A charcoal fire built that was 5._____ enough to cover the end of a sow.

To 6._____ cast iron into wrought iron, heavy pigs and sows were dragged from the furnace to the forge by oxen. They were 7._____ in the finery hearth through an aperture in the side of the chimney. Rollers guided the sows into the fire where they were slowly 8._____. Long iron bars or "ringers" were used to manipulate the melted iron. The melted iron was lifted up into the air blast over and over again until the carbon was sufficiently reduced. As the carbon content went down the melting temperature went up. Perhaps this was an 9._____ to the finer when the iron had reached the desired carbon content. The process produced more slag and it is possible that some slag may have been 10._____ added to assist in the carbon reduction process.

1. A. Typical   B. Typically   C. Typicality   D. Typify
2. A. sad   B. surprised   C. possible   D. glad
3. A. processed   B. processes   C. processing   D. process
4. A. aiming   B. aimed   C. aim   D. aims
5. A. large   B. larger   C. largest   D. largely
6. A. polishing   B. finish   C. improved   D. refine
7. A. positioned   B. positioning   C. position   D. putting
8. A. reduced   B. increased   C. change   D. melted

Unit 2   Iron Making

9. A. indicating   B. suggestion   C. indicator   D. suggest
10. A. purposive   B. deliberately   C. deliberate   D. purposeful

Task Four | The following passage has two paragraphs. Choose the correct sentence for each paragraph from the list of sentences below.

> A. The blast furnace uses coke, iron ore and limestone to produce pig iron.
> B. The blast furnace is the first step in producing steel from iron oxides.
> C. Coal is a key part of the coke-making process.
> D. The production of raw coal was slightly accelerated.
> E. The coke is cooled and screened into pieces ranging from one inch to four inches.

Steel is primarily produced using one of the following two methods: Blast Furnace or Electric Arc Furnace. 1._____ The first blast furnaces appeared in the 14th century and produced one ton per day. Even though the equipment is improved and higher production rates can be achieved, the processes inside the blast furnace remain the same. 2._____

3._____ The coal is crushed and ground into a powder and then charged into an oven where it is heated to approximately 1800°F in the absence of oxygen. As the oven is heated, the coal begins to melt so most of the volatile matter such as oil, tar, hydrogen, nitrogen and sulfur are removed. The cooked coal, called coke, is removed from the oven after 18 to 24 hours of reaction time. 4._____ The coke is a porous, hard black rock of concentrated carbon (contains 90 to 93 percent carbon), which has some ash and sulfur but compared to raw coal is very strong. The strong pieces of coke with a high energy value provide permeability, heat and gases which are required to reduce and melt the iron ore, pellets and sinter.

Part Ⅲ: Academic Skills

Academic Writing Skills(Ⅱ)*

Reading Critically and Drawing Relevant Material from Scholarly Texts to Craft Your Academic Writing

In every discipline, writing helps us learn to think critically about our own ideas and the ideas of others. Academic writing is closely linked to reading—you will rarely be asked to write about something without first being expected to read some appropriate texts. In academic writing you will almost always draw on the work of others and so it is essential that you learn to read critically and draw relevant material from other texts. So how do you develop that skill? How do you become a critical reader and draw relevant material from scholarly texts to craft essays?

1) Reading Critically

Academic reading is not a passive activity; to become a good academic reader you must approach the text as something that needs a response from you.

"Active Reading" requires a planned approach so that you can begin to grapple with the meaning in the text. When you are ready to read articles, use a method that works best for you to capture key points and issues. You might use a computer programme or pen and paper but always write as you read. Write from the beginning because as you write you are developing content which you can revise and structure later.

When you do this, you are drawing on the work and ideas of an author and it is important to integrate the work of others in an honest way by referencing the original source. Citing sources also shows you are entering the conversation already begun in the academic or professional community. Citing others will improve your academic writing by clearly creating an intellectual basis and framework for your writing and result in better writing.

Three skills are particularly important as you write: **Summarising, Paraphrasing and Direct Quotation. Summarising** means writing an overview in your own words of the main ideas, issues and general meanings of a text. It is about giving a general picture where you cite the original author. Sources help the writer to make a point and academic writers have a responsibility to cite all

---

* Taken from *Developing Your Academic Writing Skills: A Handbook* by Marian Fitzmaurice & Ciara O'Farrell.

sources used. **Paraphrasing** means focusing on a particular issue, idea or section in a text and using your own words to put forward the meaning of the original text. In a paraphrase you do not directly quote the text but, again, you must cite the original author. Try and be confident to write in your own voice and to paraphrase in your own words. **Direct Quotation** is usually identified by quotation marks or block indentation and is entirely the words of the original author which you have chosen to use. Use quotation for specific purposes:

- To present a very well-stated passage of text whose meaning would be lessened if paraphrased
- To present an idea or argument to comment on

If you use an author's specific words, you must place the words within quotation marks, or in block indentation and you must credit the source.

Activity 1: Active reading

Have a book/journal article or report in front of you that you are planning to read and do the following:

1. Underline key ideas and key terms.
2. Use lines on the margin to draw attention to an important passage.
3. At the end of a chapter or paper, sketch a simple outline of the key arguments or ideas.
4. Write a number of summary sentences to give you an overall picture of what the reading is about.

2) Examples of Drawing from Sources

A Sample Piece of Text on Academic Writing

As you begin your graduate studies it will soon become clear that there are differences between your writing in school and the writing that is expected of you in college. To succeed in college you need to write well in your academic discipline as this is still the primary way in which your learning is assessed and graded. However, academic writing is challenging and good academic writing in all disciplines requires accuracy, logical structure, attention to referencing conventions and eloquence; it is rarely achieved in one draft. In academic writing it is important to lay out the aims and extent of the content material and present it in logical order and arrive at conclusions. In academic writing a detached and objective approach is required. An academic argument appeals to logic, provides evidence in support of an intellectual position and is distinguished by the lack of an emotional appeal. In academic writing, writers always interact with each others' texts and so there will be frequent references to the ideas, thinking or research of other authors writing in this field. You must give credit to those with whom you are interacting and there are structured guidelines for referencing and citation. Also, in academic writing it is important that when a claim is made it is backed up by reasons based on some form of evidence but it is expected that the author takes a critical approach.

Example of a Summary

Fitzmaurice & O' Farrell (2013) argue that academic writing is an important skill for college students and is different from the writing they undertook in school. It is challenging as it requires the author to be objective, logical in approach, and critical; when a claim is made it must be supported by evidence. The authors point out that in academic writing it is important to draw on the ideas or research of other authors in the field and correctly reference all sources.

Example of Paraphrasing

Academic writing presents challenges as it demands clarity in presentation, a clear structure, correct referencing and fluency of expression which may not be achieved in one draft.

Direct Quotation

Fitzmaurice & O' Farrell argue that "academic writing is challenging and good academic writing in all disciplines requires accuracy, logical structure, attention to referencing conventions and eloquence; it is rarely achieved in one draft".

You have just read some examples of different ways of drawing from your sources to explain a concept or develop a point; you must cite all your sources and ensure academic integrity.

As a student in higher education it is not enough to be able to summarise, paraphrase or quote from what you have read. You need also to engage in critical reading, which requires you to think about, assess and give consideration to the texts of other scholars. The three questions outlined below, which are adapted from Wallace and Gray (2006), provide a framework for a critical reading of a text and build on the work already outlined. Use the same piece of text you were working on and write in answers to the questions.

As you do this in relation to a number of texts, you are making judgements about what different writers are saying; you will also need to adopt a critical stance.

Unit 2   Iron Making

Activity 2: A critical synopsis of a text

> Consider the following questions:
> 1. What are the authors saying that has relevance to my work?
> This question requires you to consider the links, if any, to your own project, assignment or research. Answer this question in 25 words.
> 2. How convincing are the authors' statements?
> This question requires you to evaluate the arguments put forward by the authors.
> - What claims are made?
> - Are there unsubstantiated claims?
> - What evidence is used to support the arguments? Is there a data set used and, if so, are the claims clearly related to this?
> - Are the claims consistent with other articles you have read? If not in what way do the claims differ?
>
> Write one sentence in answer to each of the above questions.
> 3. What use can I make of this in my assignment?

This question requires you to think about whether this is a key text that you will use and discuss in depth. Think about how the claims made here tie in with what you believe. Or if they can be used to expand or question other claims that you have read. Or will you only refer to it briefly? Freewrite for 5 minutes to answer this question.

Remember to feel free to disagree with the experts, explaining why you do so and that your own analysis is the star; the views of others play a supporting role.

Part IV: Extended Reading and Translation

Translate the following passage into Chinese.

Regarding economic development, there are three periods that are particularly notable. The first period starts from the foundation of the People's Republic of China in 1949, which stands out for the construction of China's industrial base, with the emphasis on the extraction of coal and the creation of industries such as steel and materials, construction, heavy engineering, chemistry, power generation, among other things.

The second period is the structural reforms starting from the late 1970s when the country introduced special economic zones, attracted foreign investment, developed foreign trade and generated great employment.

Since 2013, the country has embraced a new era that stands out for the pursuit of systemic development based on a modern governance model for China. This model is composed of themes, objectives and actions. Its basic premises are "focus on the well-being" of the Chinese people and realizing the Chinese Dream of national rejuvenation. In other words, it aims to eradicate poverty, reduce social inequalities and create better living conditions for all Chinese people.

Unit 3 Steel Making

Part Ⅰ: Text A

Section A Warm-up Questions

Discuss with your partner about the following questions.
1. Where are carbon steels used in the world?
2. What are the benefits of stainless steels?
3. Is stainless steel corrosion resistant?

Section B Listening Practice

Listen to the passage and fill in the blanks below.

People have been recycling metals for hundreds of years. Today, re-using metal waste or 1._____ provides work for many people, especially in developing countries. Three kinds of metals are recycled. They are ferrous metals, nonferrous metals and precious metals.

Ferrous metals contain iron. They are low in cost and recycled in large amounts. Metallic iron called pig iron is produced when iron is 2._____ in a hot industrial stove. This kind of stove is called a blast furnace. Pig iron 3._____, carbon. Pig iron is useful because it can be formed into solid, heavy objects or objects with unusual shapes.

Another kind of iron is steel, which is iron without the carbon. Making steel is simply removing the carbon by 4._____. This makes the steel stronger and easier to cut than iron. Both pig iron and steel waste or scrap are useful because they 5._____ to make new products.

In countries where there is 6._____ steel scrap, old tin cans are sometimes used. Tin cans are mostly steel. They can be melted. If the scrap is heated before the temperature gets to 7. _____, the blast furnace can be more simply designed and less costly. These simpler furnaces are called foundries. Products are made in foundries all over the world, but especially in Asia.

Nonferrous metals include copper and aluminum. Copper is the perfect material for recycling. It is valuable, 8. _____ and easy to clean. People who operate foundries around the world buy copper wire and cable to recycle. Aluminum is another very popular nonferrous scrap metal. It is cheap to produce and very easy to work with. In developing countries, small foundries produce aluminum bars, sheets and wires.

Precious metals like silver also are recycled. Silver can be found in pictures 9. _____. And it can be found in X-rays after they have been developed. X-ray film is 10. _____ silver, because both sides of the film are usually developed.

☞ Section C Active Reading

Standard Steels

1 Steel is the generic term for a large family of iron-carbon alloys that are malleable, within some temperature range, immediately after **solidification** from the molten state. The principal raw materials used in steelmaking are iron ore, coal, and limestone. These materials are converted in a blast furnace into a product known as "pig iron", which contains considerable amounts of carbon, manganese, sulfur, phosphorus, and silicon. Pig iron is hard, brittle, and unsuitable for direct processing into wrought forms. Steelmaking is the process of refining pig iron as well as iron and steel scrap by removing undesirable elements from the melt and then adding desirable elements in **predetermined** amounts. A primary reaction in most steelmaking is the combination of carbon with oxygen to form a gas. If **dissolved** oxygen is not removed from the melt prior to or during pouring, the **gaseous** products continue to evolve during solidification. If the steel is strongly **deoxidized** by the addition of deoxidizing elements, no gas is evolved, and the steel is called "killed" because it lies quietly in the molds. Increasing degrees of gas evolution (decreased deoxidation) characterize steels called "**semikilled**". The degree of deoxidation affects some of the properties of the steel. In addition to oxygen, liquid steel contains measurable amounts of dissolved **hydrogen** and **nitrogen**. For some critical steel applications, special deoxidation practices as well as vacuum treatments may be used to reduce and control dissolved gases.

2 The carbon content of common steel grades ranges from a few hundredths of a percent to about 1 percent. All steels also contain varying amounts of other elements, principally manganese, which

acts as a deoxidizer and facilitates hot working. Silicon, phosphorus and **sulfur** are also always present, if only in trace amounts. Other elements may be present, either as **residuals** that are not intentionally added but result from the raw materials or steel-making practice, or as alloying elements added to effect changes in the properties of the steel.

3 Steels can be cast to shape, or the cast ingot or **strand** can be reheated and hot worked by rolling, forging, extrusion, or other processes into a wrought mill shape. Wrought steels are the most widely used of engineering materials, offering a **multitude** of forms, finishes, strengths, and usable temperature ranges. No other material offers comparable versatility for product design.

4 Wrought steels may be classified systematically into groups based on some common characteristic, such as chemical composition, deoxidation practice, finishing method, or product form. Chemical composition is the most often used basis for identifying and assigning standard designations to wrought steels. Although carbon is the principal hardening and strengthening element in steel, no single element controls the steel's characteristics. The combined effect of several elements influences response to heat treatment, hardness, strength, microstructure, corrosion resistance, and **formability**. The standard steels can be divided broadly into three main groups: carbon steels, alloy steels, and stainless steels.

Carbon Steels

5 A steel qualifies as a carbon steel when its manganese content is limited to 1.65 percent (max), silicon to 0.60 percent (max), and copper to 0.60 percent (max). With the exception of deoxidizers and boron when specified, no other alloying elements are added intentionally, but they may be present as residuals. If any of these incidental elements are considered **detrimental** for special applications, maximum acceptable limits may be specified. In contrast to most alloy steels, carbon steels are most often used without a final heat treatment; however, they may be **annealed**, **normalized**, case hardened, or **quenched** and **tempered** to enhance fabrication or mechanical properties. Carbon steels may be killed, semikilled, capped, or rimmed, and, when necessary, the method of deoxidation may be specified.

Alloy Steels

6 Alloy steels comprise not only those grades that exceed the element content limits for carbon steel, but also any grade to which different elements than those used for carbon steel are added, within specific ranges or specific minimums, to enhance mechanical properties, fabricating characteristics, or any other attribute of the steel. By this definition, alloy steels **encompass** all steels other than carbon steels; however, by convention, steels containing over 3.99 percent chromium are considered "special types" of alloy steel, which include the stainless steels and many of the tool steels.

7   In a technical sense, the term alloy steel is reserved for those steels that contain a modest amount of alloying elements (about 1–4 percent) and generally depend on thermal treatments to develop specific mechanical properties. Alloy steels are always killed, but special deoxidation or melting practices, including vacuum, may be specified for special critical applications. Alloy steels generally require additional care throughout their manufacture because they are more sensitive to **thermal** and mechanical operations.

Stainless Steels

8   Stainless steels are high-alloy steels and have superior corrosion resistance to the carbon and conventional **low-alloy** steels because they contain relatively large amounts of chromium. Although other elements may also increase corrosion resistance, their usefulness in this respect is limited.

9   Stainless steels generally contain at least 10 percent chromium, with or without other elements. It has been customary in the United States, however, to include in the stainless steel classification those steels that contain as little as 4 percent chromium. Together, these steels form a family known as the stainless and heat-resisting steels, some of which possess very high strength and **oxidation** resistance. Few, however, contain more than 30 percent chromium or less than 50 percent iron.

10   In the broadest sense, the standard stainless steels can be divided into three groups based on their structures: **austenitic**, **ferritic**, and **martensitic**. In each of the three groups, there is one composition that represents the basic, general-purpose alloy. All other compositions are derived from the basic alloy, with specific variations in composition being made to obtain very specific properties.

◆ Words and Expressions

solidification /səˌlɪdɪfɪˈkeɪʃn/ n. the process of becoming hard or solid by cooling or drying or crystallization 凝固；浓缩

predetermined /ˌpriːdɪˈtɜːmɪnd/ adj. the form or nature is decided by previous events or by people rather than by chance 预先确定的

dissolve /dɪˈzɒlv/ v. to mix with a liquid and become part of it 使(固体)溶解

gaseous /ˈɡæsiəs/ adj. like or containing gas 似气体的；含气体的

deoxidize /diːˈɒksɪdaɪz/ v. to remove combined oxygen from (a substance, usually a metal) 使(金属等)脱氧

residual /rɪˈzɪdjuəl/ n. something left after other parts have been taken away 残留物；剩余

strand /strænd/ n. a single thin piece of thread, wire, hair, etc. (线、绳、金属线等的)股，缕

multitude /ˈmʌltɪtjuːd/ n. an extremely large number of things or people 众多；大量

detrimental /ˌdetrɪˈmentl/ adj. causing harm or injury 有害的；不利的

Unit 3 Steel Making

anneal / ə'niːl/ *v.* to temper or toughen (something) by heat treatment 使退火

normalize / 'nɔːməlaɪz/ *v.* to fit or make something fit a normal pattern or condition （使）正常化，标准化

quench / kwentʃ/ *v.* to stop a fire from burning 扑灭；熄灭

temper / 'tempə(r)/ *v.* to make metal as hard as it needs to be by heating and then cooling it 使（金属）回火

encompass / ɪn'kʌmpəs/ *v.* to include a large number or range of things 包括，涉及（大量事物）

thermal / 'θɜːml/ *adj.* connected with heat 热的；热量的

oxidation / ˌɒksɪ'deɪʃn/ *n.* a process in which a chemical substance changes because of the addition of oxygen 氧化

◆ Terminology

semikilled / 'seməkɪld/ *adj.* 半镇静钢的（半脱氧钢的）

hydrogen / 'haɪdrədʒən/ *n.* 氢；氢气

nitrogen / 'naɪtrədʒən/ *n.* 氮；氮气

sulfur / 'sʌlfə(r)/ *n.* 硫

formability / ˌfɔːmə'bɪlɪti/ *n.* 成型性，成型性能

low-alloy / 'ləuˌælɔi/ *adj.* 低合金的

austenitic / ˌɔːstə'nɪtɪk/ *adj.* 奥氏体的

ferritic / fə'rɪtɪk/ *adj.* 铁素体的，铁氧体的

martensitic / ˌmɑːtin'zitik/ *adj.* 马氏体的

☞ Section D Language Focus

Task One | Text organization

Work in groups and discuss the organization of the text and fill in the blanks.

Parts	Paragraphs	Main Ideas
Part 1	Para. 1–2	1.
Part 2	Para. 3–4	2.
Part 3	Para. 5–10	3.

Task Two | Answer the following questions based on the information contained in the text.

1. Which of the following is not the main raw material for steelmaking? _____

   A. Iron.

   B. Ore.

   C. Steel.

   D. Coal.

2. The word "deoxidized" in the first paragraph means _____.

   A. traced the course

   B. became less in quantity

   C. removed oxygen from a compound

   D. made it smaller in size or amount

3. Which is the most widely used of engineering materials? _____

   A. Wrought steels.

   B. Alloy steels.

   C. Carbon steels.

   D. Stainless steels.

4. Which of the following is an accurate description of carbon steels? _____

   A. They require additional care throughout their manufacture.

   B. They contain at least 10 percent chromium.

   C. They are most often used without a final heat treatment.

   D. They are more sensitive to thermal and mechanical operations.

5. From this passage, we can learn that _____.

   A. it is necessary to lower the amount of carbon dioxide

   B. steelmaking has been regarded as a profitable business

   C. standard steels have their own properties and can be divided into three principal types

   D. carbon steels have superior corrosion resistance to stainless steels

Task Three | Fill in the blanks with the words or phrases given in the box. You may not use any of the words or phrases more than once.

Stainless steels are iron-based alloys. High 1. _____ to corrosion is the principal 2. _____ of stainless steels, which is provided by chromium—the main alloying element. Enhanced corrosion resistance may be imparted by additional alloying 3. _____, primarily nickel and molybdenum.

Unit 3   Steel Making

Stainless steels have a wide 4. _____ of applications, including those where human health may be involved. In these 5. _____, stainless steels may come into direct contact with the human body. Cases 6. _____ jewellery, cutlery, medical devices, automotive applications, food preparation and pharmaceutical settings. Stainless steel can be used without 7. _____ any negative influence on food, drinking water, beverages, or medical preparations it is in contact with.

Their health and hygiene characteristics—for instance, being easier to disinfect or clean—mean stainless steels 8. _____ to human health and to society—ensuring that it is possible to safely prepare and store food and beverages, conduct medical interventions or to purify and channel water. 9. _____, stainless steel has excellent environmental characteristics. It does not seep into water and is safe in 10. _____ with organisms.

A. range	I. Additionally
B. include	J. constituents
C. contact	K. isolate
D. resistance	L. reactive
E. elements	M. applications
F. exerting	N. contribute
G. release	O. environment
H. characteristic	

Task Four | The following passage has three paragraphs. Choose the correct clause for each paragraph from the list of clauses below.

A. due to its unique qualities that deliver strength and durability along with low density

B. an industry that requires lower density in addition to strength

C. when money is an important part of the equation

D. differences existing between titanium and stainless steel

E. while titanium's characteristics are naturally found within it

The main difference between stainless steel and titanium is simply that stainless steel is an alloy metal while titanium is a metal. The unique characteristics of stainless steel are created by adding alloying metals to it, 1. _____.

Circumstances exist that often suggest one substance is better suited than the other for use in a specific project or activity. For example, titanium is often preferred by some manufacturers 2. _____. Therefore, when weight is a more important consideration than strength, titanium is often preferred. Conversely, stainless steel is preferred by industries that place a higher importance on weight than strength. While titanium is not as dense as steel, it is just as strong, making it highly suitable for specific industries, such as aerospace, 3. _____.

Titanium is more expensive, though, than stainless steel, making it cost-prohibitive for some industries, such as construction, which requires large quantities. Therefore, 4. _____, stainless steel is sometimes chosen over titanium if both substances are deemed suitable.

Task Five | Translate the following sentences into English, using the words or phrases in the brackets.

1. 该证件将在预定年限后到期。(predetermined)

2. 无论你是否加热，这种物质不溶于水。(dissolve)

3. 一些食物因所含的化学成分被怀疑有害健康。(detrimental)

4. 这座城堡被一条护城河围绕着。(encompass)

5. 有足够的水供他们解渴。(quench)

Part Ⅱ: Text B

☞ Section A Warm-up Questions

Discuss with your partner about the following questions.
1. How is the slag used in our daily life?
2. What is the difference between a slag and a melted metal?
3. Why is it important to recycle slag by-products?

☞ Section B Listening Practice

Listen to the passage and fill in the blanks below.

Moon Base Work Yields Clean Steel Process

Flaming cauldrons of molten metal have long been the primary venues for steel production. But blast furnaces require a lot of coal, which means greenhouse gas 1._____. In fact, worldwide, steelmaking is 2._____ for 5 percent of annual emissions.

But scientists working on a way to 3._____ oxygen from the iron oxide in lunar soil for future moon bases realized that they happened on a better way to make steel here on Earth. The 4._____? Produce steel the way we make aluminum: use electricity rather than 5._____.

To make steel the 6._____ way, you blast iron ore with heat and purify the resulting molten metal with oxygen. The process 7._____ carbon from the steel, but produces carbon dioxide. Making a ton of steel 8._____ roughly two tons of CO_2—and the world uses a lot of steel in cars, buildings and other infrastructure.

The new method 9._____ passing a current through a molten pool of iron oxide, which drives off the originally sought-after oxygen. The by-product is steel. And depending on the source of the electricity, the process could be nearly CO_2-free. Which, as far as the 10._____ is concerned, would be very cool.

Section C Active Reading

An Overview of Steelmaking Slag

1   Vast amounts of slags and other by-products are produced as an output of steel and iron production process. Limitations as to the storage of solid wastes and other environmental issues have **rendered** the recycling of such waste materials a highly critical subject for protection of natural sources. Beside such environmental **concerns**, utilization of industrial wastes in industrial applications for energy and cost saving purposes, and their recycling into **applicable** by-products has been the subject of various attempts in recent decades. Considering the local and global steel production, the need for utilization of slag as a value added by-product or as extra energy output, becomes both a requirement for environmental concerns and an opportunity for cost-saving in industrial applications. Industrial slags have found usage in some of the low-value added sectors as well. These applications include their use as filling material in road construction, as **additive** in cement production, in production of railway **ballasts**, in glass production and in production of thermal **insulation** wool.

2   Blast furnace slags are mainly composed of **ceramic** based compounds. Such oxide contents render blast furnace slags **candidate** materials for thermal insulation purposes due to their significantly lower **coefficients** of thermal **conductivity** as compared to metallic materials. Recently, such characteristics of slags have drawn the attention of researchers to their use against severe conditions such as oxidation, **corrosion** and thermal cycles through their **deposition** or application on target surfaces by means of surface engineering techniques, thus paving the way for their utilization in value added applications as well.

3   The mentioned properties of industrial slag bring about the question of whether this waste material can be utilized in value-added applications as well, in addition to its use in low-value added applications such as filling material in construction industry or for waste heat recovery in steel production plants for cost-saving. In this regard, the current and possible use of industrial slag in the fields such as surface engineering is yet to be investigated through **elaboration** of literature works related to the **composition**, characteristics and utilization of various slag types.

4   In general terms, slag is a waste material of the steelmaking process which is composed of a mixture of metal oxides, SiO_2, metal **sulfides** and elemental metals that accumulate on the surface of impure molten metals. Metals in raw form are generally referred to as ores since in nature they exist in the form of oxides, sulfides, and other compounds or as a mixture of these metallic and non-

metallic compounds. Reduction of ores in chemical compound form into the pure metal form requires inclusion of **flux** materials which bond with impurities for formation and removal of slag from molten metal. Flux materials such as **dolomite** and **burnt lime** are commonly used for this purpose. The slag layer that **accumulate** on high temperature molten metal during the melting of ores is removed from the surface and solidified via various cooling methods.

5 **Reutilization** of industrial waste slags as industrial by-products through their recycling depends greatly on the slag properties and the heat treatment that the molten metal has undergone. In blast furnaces (BF) ore **gangue** and non-metallic impurities are removed from the metal ore, thereby producing pig iron and slag as the final products. Pig iron and slag accumulate at the center of the BF in molten form, then slag **diffuses** to the molten metal surface due to its lower density than iron. The high **viscosity** of slag, which reaches even higher values at low temperature regions of furnace requires application of additional processes for its removal from BF. When removed from the furnace, slag is still in molten form with high temperature. The operations performed to cool down and solidify molten slag for storage become effective on slag composition.

6 Solidification of slag from high temperatures with suitable cooling processes holds critical importance for improvement of its properties, and its applicability in recycling applications. Two types of solid **phases** are observed during solidification of molten slag. One type is crystal phase which results from slow cooling of molten slag, another type is **amorphous** or **vitreous** phase which forms as a result of fast cooling processes (such as quenching). These two slag types in amorphous and **crystal** form have different fields of use. Slags with amorphous (or vitreous) phase, which are the products of fast solidification, are generally used in applications other than recovery of waste heat. BFSs are also used for acquisition of extra melting heat in blast furnaces, and the waste heat recovery of slag is highly dependent on the speed of the cooling process. In this regard, the speed of the cooling process can be either slow or fast depending on the intended use of the solidified slag (waste heat recovery or better recycling properties).

◆ Words and Expressions

render / ˈrendə(r)/ v. to cause somebody/something to be in a particular state or condition 使成为；使处于某状态

concern / kənˈsɜːn/ n. a feeling of worry, especially one that is shared by many people（尤指许多人共同的）担心，忧虑

applicable / əˈplɪkəbl/ adj. relevant to a particular situation or can be applied to it 适用的

additive / ˈædətɪv/ n. a substance that is added in small amounts to something, especially food, in order to improve it, give it colour, make it last longer, etc. (尤指食品的)添加剂, 添加物

insulation /ˌɪnsjuˈleɪʃn/ n. the act of protecting something with a material that prevents heat, sound, electricity, etc. from passing through 隔热; 隔音; 绝缘

ceramic /səˈræmɪk/ n. a pot or other object made of clay that has been made permanently hard by heat 陶瓷制品; 陶瓷器

candidate /ˈkændɪdət/ n. a person or thing that is regarded as being suitable for a particular purpose or as being likely to do or be a particular thing 有望做……的人; 有望成为……的事

coefficient /ˌkəʊɪˈfɪʃnt/ n. a number that expresses a measurement of a particular quality of a substance or object under specified conditions 系数

corrosion /kəˈrəʊʒən/ n. the damage that is caused when something is corroded 腐蚀

deposition /ˌdepəˈzɪʃn/ n. the natural process of leaving a layer of a substance on rocks or soil; a substance left in this way 沉积(物); 沉淀(物)

elaboration /ɪˌlæbəˈreɪʃn/ n. an explanation or description of something in a more detailed way 详尽阐述

composition /ˌkɒmpəˈzɪʃn/ n. the different parts which sth. is made of; the way in which the different parts are organized 成分; 构成

accumulate /əˈkjuːmjəleɪt/ v. to gradually get more and more of something over a period of time 积累; 积聚

diffuse /dɪˈfjuːz/ v. if a gas or liquid diffuses or is diffused in a substance, it becomes slowly mixed with that substance (使气体或液体)扩散, 弥漫, 渗透

phase /feɪz/ n. a stage in a process of change or development 阶段; 时期

amorphous /əˈmɔːfəs/ adj. having no definite shape, form or structure 不规则的

vitreous /ˈvɪtriəs/ adj. hard, shiny and transparent like glass 玻璃质的; 透明的

◆ **Terminology**

ballast /ˈbæləst/ n. (用作公路或铁路路基的)道砟

conductivity /ˌkɒndʌkˈtɪvəti/ n. 导电性

sulfide /ˈsʌlfaɪd/ n. 硫化物

flux /flʌks/ n. 通量; 流动

dolomite /ˈdɒləmaɪt/ n. 白云石

burnt lime 煅石灰; 氧化钙

reutilization /ˌriːjuːtəlaɪˈzeɪʃn/ *n.* 二次利用；重复利用

gangue /gæŋ/ *n.* 脉石

viscosity /vɪˈskɒsəti/ *n.* 黏性；黏度

crystal /ˈkrɪstl/ *adj.* 透明的；晶体的

☞ Section D Language Focus

Task One | Are the following statements True or False according to the passage? Write T/F accordingly.

1. (   ) A lot of slags and other by-products are produced in the process of iron and steel production.

2. (   ) The use and recycling of industrial wastes in industrial applications is the theme of various attempts in the past decade.

3. (   ) Slag is an industrial by-product obtained from the steel manufacturing industry.

4. (   ) When removed from the furnace, slag can not stay in molten form at high temperature.

5. (   ) It is very important to adopt proper cooling process to solidify slag for improving the properties of slag and its applicability in recycling.

Task Two | Translate the following sentences into Chinese.

1. Limitations as to the storage of solid wastes and other environmental issues have rendered the recycling of such waste materials a highly critical subject for protection of natural sources.

2. Recently, such characteristics of slags have drawn the attention of researchers to their use against severe conditions.

Unit 3 Steel Making

3. In this regard, the current and possible use of industrial slag in the fields such as surface engineering is yet to be investigated through elaboration of literature works related to the composition, characteristics and utilization of various slag types.

4. In general terms, slag is a waste material of the steelmaking process.

5. The operations performed to cool down and solidify molten slag for storage become effective on slag composition.

Task Three | Choose the best answers for each blank of the following passage.

Slag is a by-product of metal smelting, and hundreds of tons of it are produced every year all over the world in the process of 1. _____ metals and making alloys. Like other industrial by-products, slag actually has many uses, and 2. _____ goes to waste. It appears in concrete, aggregate road materials, as ballast, and is sometimes used as a 3. _____ of phosphate fertilizer. In 4. _____, slag looks like a loose collection of aggregate, with lumps of varying sizes. It is also sometimes 5. _____ to as cinder, in a reference to its sometimes dark and crumbly appearance.

This substance is produced during the 6. _____ process in several ways. Firstly, slag represents 7. _____ impurities in the metals, which float to the top during the smelting process. Secondly, metals start to oxidize as they are smelted, and slag forms a 8. _____ crust of oxides on the top of the metal being smelted, protecting the liquid metal underneath. When the metal is smelted to satisfaction, the slag is skimmed from the top and 9. _____ of in a slag heap to age. Aging material is an important part of the process, as it needs to be 10. _____ to the weather and allowed to break down slightly before it can be used.

1. A. refine   B. refined   C. refining   D. refines
2. A. rarely   B. rare   C. uncommon   D. unusual
3. A. essential   B. component   C. basic   D. integral
4. A. image   B. occurrence   C. development   D. appearance
5. A. referred   B. refer   C. refers   D. reference
6. A. smell   B. smelting   C. smelt   D. heat
7. A. undesired   B. desire   C. wish   D. impulse
8. A. save   B. protective   C. defend   D. protect
9. A. incline   B. encourage   C. disposed   D. persuade
10. A. reveal   B. show   C. exposed   D. uncover

Task Four | The following passage has three paragraphs. Choose the correct sentence for each paragraph from the list of sentences below.

A. It has the potential to become a significant source of landfill waste and pollution.
B. Second, care must be taken to improve slag potential for other purposes without compromising the quality of the metal it is originally used to produce.
C. One of the most common is as a constituent of concrete and cement.
D. Slag has a double role: it permits removal of impurities.
E. The goals of slag recycling efforts are both environmental and economic.

Slag is a by-product of the metallurgical smelting process. 1. _____ It also allows exchange reactions with the liquid metal, permitting control of the process in order for the desirable elements to stay in the melt while the others are removed.

Slag is generated in large quantities. 2. _____ But slag itself has value, and technologies have emerged to recycle and reuse reprocessed, granulated slag in different building materials, such as cement, brick, concrete aggregates, wall materials, and glass-ceramic tiles.

3. _____ Reprocessed slag can replace other more costly ingredients used to make building materials. Recycling slag minimizes waste and disposal costs, reduces energy use, extends furnace life, and lowers the amount of additives needed to make slag. The concerns with slag recycling are that first, because slag can contain environmentally hazardous materials, it must be analyzed for elemental composition before reuse. 4. _____ Better slag makes for a better finished product, so quick and reproducible analyses of all the oxides in the slag are needed to control slag quality as well as alloy quality during the smelting process.

Unit 3   Steel Making

 Part Ⅲ: Academic Skills

Academic Writing Skills(Ⅲ) *

1. Being Critical

Being critical is not just about praising or tearing apart the work of others. Adopting a critical stance to a text means paying close attention to the text in terms of definitions, ideas, assumptions and findings or arguments. It is focused questioning and interrogation which is respectful of what others have done and contributed. It is not about being negative about the work of others but assessing the contribution of other scholars. Asking and answering the questions which follow will help you as a student to judge the work of other scholars.

Activity 1

1. What is the argument?
2. What aspect of the topic/argument is spoken about in this article? What evidence is used to back up the argument?
3. What claims are made by the author?

As you answer these questions you are moving beyond summaries and into evaluating and becoming critical.

2. Intertextuality: Making Connections Between Texts and Putting Forward your Own Understanding

When writing your paper you will read a number of texts; the next stage is to move between the texts and draw ideas together before putting forward your own understanding. This is "Intertexual" work, an important part of academic writing, where an important task is negotiating how to relate and make connections between the ideas drawn from different writers and putting forwardyour own understanding. Almost every word and phrase we use we have heard or read before. So as we create our texts we are influenced by words or ideas already written. Intertextuality means working with a

---

* Taken from *Developing Your Academic Writing Skills: A Handbook* by Marian Fitzmaurice & Ciara O'Farrell.

number of texts and relating one text to another. Firstly, it is about drawing on other texts to build a context. It also requires you to think about how to use these texts to inform your argument and make your own assertion. As a new student to university there is a real challenge in deciding how ideas and information are joined, structured and supported. As you work through the activities in this handbook you are involved in focused questioning and examination of a number of texts which will help you to make connections between the texts, and recognise and distinguish the major ideas, arguments and debates about a topic. This is what intertextual work is all about. As you are seeking to analyse relationships among sources it is useful to have a list of phrases which you can incorporate into your own work such as: According to... or X argues for... Others have suggested... Y has shown in her study... In his article Z concludes.

Activity 2

Find two scholarly articles from your own discipline and answer the following questions:

1. How does the writer create a context using the texts of others?
2. How does the writer use the texts of others to build his/her own argument?
3. Write down any connections or differences you can see between the two articles.

Unit 3 Steel Making

Part Ⅳ: Extended Reading and Translation

Translate the following passage into Chinese.

Iron ore is the most important raw material in steel production. Its resources are found mainly in Australia, Brazil, China, India, Russia and the United States. Australia, China, Brazil and India alone, account for around 80 percent of the world total output.

According to statistics from the Metallurgical Mines' Association of China, global iron ore consumption in 2001 was over 1 billion tons. In 2010, it nearly doubled to more than 1.8 billion tons.

China contributed to 90 percent of the increase. The country has become the world's largest iron ore consumer, and the world's second-largest producer. But China's iron ore resources are scattered, of low grade and with a high cost of extraction. It relies highly on imports from other countries.

Last year, it reported 440 million tons in output, 740 million tons in imports and 1.05 billion tons in consumption. Most of the imports are from Brazil's Vale, Australia's Rio Tinto, BHP and FMG.

Unit 4　Introduction to Nonferrous Metallurgy

Part Ⅰ：Text A

Section A　Warm-up Questions

Discuss with your partner about the following questions.

1. Do you know anything about nonferrous metals? Could you give some examples of them?
2. What are the common properties of nonferrous metals?

Section B　Listening Practice

Watch the video clip and answer the following questions.

1. What is the difference between ferrous and nonferrous metals?
2. When did ferrous metals come into existence?
3. What are the properties of most ferrous metals mentioned in the video clip?
4. What is the main advantage of most ferrous metals mentioned in the video clip?
5. Whether stainless steel is a nonferrous metal or a ferrous metal?

Section C　Active Reading

Introduction to Nonferrous Metallurgy

1　Iron and steel have mechanical properties useful for many applications. When steel is

Unit 4 Introduction to Nonferrous Metallurgy

manufactured from iron, its composition can be changed through a number of metallurgical processes to **attain** different characteristics. Ferrous metallurgy, the science of developing metals that use iron as their major alloying element, has been the main focus of this passage.

2 As versatile as iron and steel are, however, there are many types of nonferrous metals and alloys that are widely used by metallurgists. Nonferrous metallurgy is the study of metals that do not use iron as their principal alloying element. The basic properties of nonferrous metals (such as copper, aluminum, and **titanium**) make them well suited for many applications.

3 Can you imagine an airplane made entirely of steel? It would be impossible for the aircraft to carry much **payload**, even if it could take off! Such an application would require lighter metals for manufacture. Light weight and other special properties can be attained through the application of nonferrous metallurgy and alloys.

4 Often, the basic methods for improving the properties of nonferrous metals are similar to the metallurgical processes used for steel. This paper will introduce some of these methods and discuss one of them in detail.

5 Most pure metals are quite soft by nature. They will bend easily and they will **wear out** if rubbed against other hard materials for a period of time. If you **grind** a sharp edge into a piece of pure metal to make a knife, for example, the edge becomes dull very quickly after use.

6 As we know, the use of alloys and different heat treatment processes can improve the hardness and strength of iron. It is often necessary to improve similar characteristics in nonferrous metals. The mechanical properties of pure nonferrous metals can be changed using three basic processing methods:

- Alloying
- Cold working and **annealing**.
- **Precipitation** hardening.

7 One of these methods will be discussed then. The metallurgical processes used for nonferrous alloys are very similar to those used for some types of alloy steel, especially stainless steel.

8 Nonferrous metallurgy can be used to produce other desirable properties besides hardness and strength. Some applications require not only tensile strength or hardness, but also **ductility** and formability. Other special properties offered by nonferrous metals include ease of processing, lighter weight, corrosion resistance, and electrical conductivity.

9 Precipitation hardening is a heat treatment process that strengthens alloys by causing phases (regions in the metal with different crystal structures) to precipitate at various temperatures and cooling rates. The metal is changed from a solution to a mixture. Suppose additional elements are added to a pure metal to form an alloy. When this alloy is heated to a high temperature, the alloying

elements form a solution in the metal. Upon rapid quenching, a new phase will precipitate in a few minutes or weeks. The new phase strengthens the metal as it grows, in this process called **age hardening**. It is used for many types of nonferrous alloys, such as copper-**beryllium**, copper-aluminum, and some stainless steel.

10 Precipitation hardening consists of two separate heating and cooling cycles. First, the alloy is heated to a **solutionizing temperature** (an elevated temperature below the melting point). In this condition, the alloy is a single-phase (or single-crystal structure) solid solution. The alloy is held at this temperature for a time before rapid quenching. Quenching prevents a second phase (or additional phases) from precipitating in the alloy. It also strengthens the metal somewhat.

11 In the second stage of precipitation hardening, the alloy is reheated to a moderate temperature (well below the solutionizing temperature), causing the second phase to precipitate. This allows the second phase component to be uniformly distributed in the grains of the original solution. The resulting structure is very strong and has the properties that are essential to the alloy.

12 The two different stages of heat treatment in precipitation hardening allow nonferrous alloys to attain maximum hardness and other desirable characteristics, such as toughness.

13 Precipitation hardening is commonly used to process copper alloys and other nonferrous metals for commercial use. The following is an example of using precipitation hardening to process C17200, a copper alloy with 1.9% beryllium, for an electrical application.

14 **Assume** a thin strip of copper-beryllium measuring 0.020″ thick and 0.3″ wide is used to make the electrical contact components in a computer. The metal is formed into contact arms by a **stamping machine**. In many cases, these components must perform two functions. The arms put force on a **pin** to make the electrical connection while also holding the pin in place. If you look closely at the socket of a **ribbon connector** in a personal computer, you may be able to see a similar application. The sockets for edge card connections in industrial computers are typically made in the same fashion.

15 In the form received at the stamping machine, C17200 is fully annealed and very soft. It also has high formability. This allows the stamping machine to make sharp bends in the metal and produce the required shape without cracking. However, the strength of this soft metal is not suitable for the computer application. The metal cannot press firmly enough against the electrical connection points.

16 Precipitation hardening can be used to produce the required strength after the metal is formed to the desired shape. In the first stage of heat treatment, the copper-beryllium parts are heated to a solutionizing temperature and held at that temperature for about 30 minutes. The parts are then rapidly quenched in water. This makes the alloy harder and stronger than it was originally.

However, the parts are not yet at full strength.

17 In some cases, the quenched parts are naturally aged at room temperature. In this step, the metal gains strength slowly over a period of time, sometimes years. For some commercial applications, however, processing must occur rapidly and the metal must have the absolute maximum strength.

18 To acquire the maximum strength in precipitation hardening, the solutionized and quenched parts are artificially aged at moderate temperature. This form of secondary heat treatment takes one to four hours. After the parts are cooled, they are at maximum strength. In some nonferrous alloys, such as aluminum A96061, some ductility is **retained** after precipitation hardening. By contrast, C17200 has very high strength but little ductility.

19 After processing, the copper-beryllium parts are strong enough to perform as electrical connectors. Solutionizing (the initial heat treatment) gave the metal excellent formability. Precipitation hardening strengthened the metal and made it suitable for commercial use.

20 Heating temperatures in precipitation hardening must be carefully controlled to produce the desired results. During heat treatment, different atomic structures, or phases, are present in the alloy at various temperatures. These phases have a direct effect on the properties of the metal and can be **identified** at each stage of heating and cooling. Through the use of phase diagrams, metallurgists can **predict** the phases that occur during precipitation hardening.

◆ Words and Expressions

attain / əˈteɪn/ v. to succeed in getting something, usually after a lot of effort (通常经过努力)获得；得到

payload / ˈpeɪləʊd/ n. the goods that a vehicle, for example a lorry/truck, is carrying; the amount it is carrying (车辆等的)装载货物；装载量

grind / graɪnd/ v. to make something sharp or smooth by rubbing it against a hard surface 使锋利；磨快；磨光

ductility / dʌkˈtɪləti/ n. the ability of a metal to be easily bent or stretched 延展性；柔软性

assume / əˈsjuːm/ v. to think or accept that something is true but without having proof of it 假定；假设；认为

pin / pɪn/ n. one of the metal parts that stick out of an electric plug and fit into a socket (插头的)销

ribbon / ˈrɪbən/ n. something that is long and narrow in shape 带状物；狭长的东西

connector / kəˈnektə(r)/ n. a thing that links two or more things together 连接物；连接器；

连线

retain /rɪˈteɪn/ *v.* to keep something; to continue to have something 保持；持有；保留；继续拥有

identify /aɪˈdentɪfaɪ/ *v.* to recognize somebody/something and be able to say who or what they are 确认；认出；鉴定

predict /prɪˈdɪkt/ *v.* to say that something will happen in the future 预言；预告；预报

wear out to become, or make something become, thin or no longer able to be used, usually because it has been used too much 磨薄；穿破；磨损；用坏

◆ Terminology

titanium /tɪˈteɪniəm/ *n.* (symbol Ti) 钛
anneal /əˈniːl/ *v.* 给（金属或玻璃）退火
precipitation /prɪˌsɪpɪˈteɪʃən/ *n.* 沉淀；淀析
beryllium /bəˈrɪliəm/ *n.* 铍
age hardening 时效硬化
solutionizing temperature 固溶温度
stamp machine 压印机；冲床；烫金机；锤击机

☞ Section D　Language Focus

Task One │ Text organization

Work in groups and discuss the organization of the text and fill in the blanks.

Parts	Paragraphs	Main Ideas
Part 1	Para. 1–4	Brief introduction to 1.＿＿＿＿
Part 2	Para. 5–8	Basic 2.＿＿＿＿ methods used in nonferrous metallurgy
Part 3	Para. 9–12	Introduction to 3.＿＿＿＿ hardening
Part 4	Para. 13–20	Using 4.＿＿＿＿ hardening to strengthen 5.＿＿＿＿

Unit 4 Introduction to Nonferrous Metallurgy

Task Two | Answer the following questions based on the information contained in the text.

1. What does "such an application" refer to in Paragraph 3? _____
   A. Steel.
   B. Aircraft.
   C. Payload.
   D. Alloy.

2. What property of iron can be improved when using alloys and different heat treatment processes? _____
   A. Softness.
   B. Flexibility.
   C. Strength.
   D. Rust.

3. Which statement about precipitation hardening is NOT true? _____
   A. Precipitation hardening is a heat treatment process that helps make alloys stronger.
   B. Precipitation hardening involves heating a mixture to a high temperature, then cooling, and finally heating to a medium temperature.
   C. In the first stage of precipitation hardening, the alloy is heated to an elevated temperature below the melting point.
   D. Precipitation hardening is commonly used to process copper alloys and other nonferrous metals for business purpose.

4. What does the word "aged" mean in the first sentence of Paragraph 18? _____
   A. Strengthened.
   B. Very old.
   C. Analyzed.
   D. Heated.

5. How can metallurgists predict the phases that occur during precipitation hardening? _____
   A. Through the use of temperature analysis.
   B. Through the use of cooling.
   C. Through the use of heating.
   D. Through the use of phase diagrams.

Task Three | Fill in the blanks with the words or phrases given in the box. You may not use any of the words or phrases more than once.

One way to harden a material is by 1._____ something that blocks or slows down the 2.

_____ of dislocations. Precipitates can do just that. Precipitation hardening is the hardening of a material due to the growth of precipitates that 3._____ dislocation motion.

It is typically performed in a vacuum, inert atmosphere at temperatures 4._____ from between 900 degrees and 1150 degrees Fahrenheit. The process ranges in time from one to several 5._____, depending on the exact material and characteristics. As with tempering, those who 6._____ precipitation hardening must strike a 7._____ between the resulting increase in strength and the loss of 8._____ and toughness. Additionally, they must be 9._____ not to over-age the material by tempering it for too long. That could result in large, spread out, and 10._____ precipitates.

A. movement	I. ineffective
B. perform	J. promote
C. hours	K. ranging
D. ductility	L. impede
E. minutes	M. balance
F. reaching	N. difference
G. adding	O. careful
H. efficient	

Task Four | The following passage has four paragraphs. Choose the correct topic sentence for each paragraph from the list of sentences below.

A. Just as water vapor in the atmosphere can condense to form liquid water, a solid dissolved in a liquid can condense to form a solid.
B. Precipitation also happens in solids.
C. More generally, precipitates are small impurity regions that form in a material when it is no longer able to dissolve the impurity.
D. Say you stir sugar into boiling water.
E. When you hear the word "precipitation", you might think of the weather.

1. _____ Precipitation occurs when water vapor in the atmosphere condenses and falls to the ground as rain, snow, sleet, or hail.

Unit 4　Introduction to Nonferrous Metallurgy

2._____This process, called precipitation, is the opposite of the process of dissolving (dissolution). A solid can dissolve in a liquid, but it can also do the opposite and "emerge" from that liquid as a solid.

3._____The sugar will dissolve in the water. So what happens when boiling water containing the maximum amount of dissolved sugar is cooled to room temperature? Since water at room temperature cannot dissolve as much sugar as boiling water, the extra sugar precipitates, or emerges as a solid from the liquid.

4._____For example, aluminum can dissolve some copper at high temperatures. When the temperature is decreased, the extra copper precipitates by forming small copper-rich regions in the aluminum. These small copper-rich regions are called precipitates.

Task Five | Translate the following sentences into English, using the words or phrases in the brackets.

1. 将刀放在一块粗石上磨，就能磨锋利。(grind)

2. 通货膨胀率预计将继续下降。(predict)

3. 可以有把握地认为火星上是没有动物生命的。(assume)

4. 公司的创始人保留了大量股份，使他迅速成为亿万富翁。(retain)

5. 你要是这么糟蹋你的外套，它很快就不能穿了。(wear out)

Unit 4　Introduction to Nonferrous Metallurgy

Part II: Text B

Section A Warm-up Questions

Discuss with your partners about the following questions.
1. Are the nonferrous environmentally friendly?
2. What do you think of the importance of nonferrous metal recycling?

Section B Listening Practice

Listen to the audio clips and fill in the blanks with missing words or phrases.

Recycling, or re-using, metals is much less costly than making them from 1. _____ in the Earth, called "ore". But when old cars, household appliances and industrial equipment are 2. _____, metals are mixed and often difficult to separate. So they are usually placed into landfills without being separated.

The United States Environmental Protection Agency says almost 3. _____ tons of this "scrap" metal is placed into landfills every year.

Although iron and steel are easily separated with 4. _____, other metals, such as copper, aluminum and titanium, are not. So products with a mix of these metals are either sent to landfills or to other countries, where workers separate the different metals 5. _____. But university scientists working with a private company say they have created a very 6. _____ to separate these light metals, using a machine.

Don Eggert is the founder of O2M Technologies. His company worked with the University of Utah to develop 7. _____. Mr. Eggert says the system is based on the knowledge that all metals 8. _____ a strong magnetic field.

"When the metal falls through the field, even (if/though) it's 9. _____, it's not attracted to the magnet, but the magnet causes there to be 10. _____ inside the metal, and that causes the metal to have a magnetic field itself, which interacts with the magnetic field that it's falling through and pushes it to the side."

Section C Active Reading

Methods of Nonferrous Metal Recycling

1 In a world which is increasingly demanding sustainability, nonferrous metal recycling has become a very important practice. Opting for recycling does not only mean being responsible towards the environment and reducing the carbon footprint, it is also a very reasonable business in industries that rely on using non-renewable resources.

2 Fortunately, nonferrous metals, especially nickel, silver, copper, aluminum and tin all share the property of being able to undergo unlimited number of instances of recycling without losing any of their original properties. That is why in Brazil, for example, as much as 98.2% of aluminum cans are recycled every year (making Brazil the number one country in that aspect) and almost 40% of all aluminum used across many different industries in the US comes from scrap.

3 Recycling is also an economically sound decision as using already extracted materials, even for metals such as aluminum which is pretty much **omnipresent**, is cheaper than mining all over again. In fact, the numbers show that in some cases it is even possible to save up to 95% of energy costs if recycling is chosen as the primary production method for aluminum.

4 However, recycling nonferrous metals can be a problematic task as they will not always come in their pure form. Often times, they are found in all sorts of liquid and solid mixtures from which they need to be extracted and purified before further use. Three methods used today for nonferrous metal recycling are electrowinning, precipitation, and nonferrous sensors.

Electrowinning

5 Electrowinning, which is also known as electroextraction, is, on the surface of it at least, a relatively simple process of extracting dissolved metals from their dissolved states using electricity. In case of nonferrous metal extraction for the purpose of recycling, the process generally goes as follows. First, the material, which can be any form of waste such as solid materials from landfills or different types of solid mixtures containing nonferrous metals, is put into a liquid solution where it is dissolved into a liquid state through the process known as leaching. The end-result of this process is called a **leachate** or leach solution. Then, using an **anode** and **cathode**—which are **electrodes** through which the current flows and which are **submerged** into the solution—an electric current is passed through the leach solution which then causes the metal(s) to be (chemically) reduced resulting in them forming a thin even layer across the surface of the submerged cathode. This way, the nonferrous metals, such as copper, tin, nickel, or silver are recovered and made readily

available for further reuse.

Precipitation

6 The second processing method for nonferrous metals is precipitation. It is also the most widely used method for metal recovery from **aqueous** solutions. Precipitation can also be used for wastewater treatment; a process in which metals are recovered from aqueous waste solutions.

7 This method includes two metal removal sub-methods called co-precipitation and **adsorption**. So as not to go into too much technical details, we will mainly address the basic method of precipitation without going too much into other of its aspects. Precipitation is the process of forming of an insoluble solid material from what originally was an aqueous solution typically involving pH adjustment or addition of another chemical species.

8 The end-result is called the "precipitate" while the chemical that causes this is called the "**precipitant**". The most commonly used precipitants are sodium and calcium hydroxides or oxides which are used to increase the pH resulting in insoluble metal hydroxides.

Metal Sensors

9 Finally, nonferrous metal sensors are becoming widely used in sorting and extracting nonferrous metals from scrap, most of which originated from end-of-life-vehicles or from e-waste. For example, sensors are used for detection and extraction of specific nonferrous metals from Zorba.

10 According to the Institute for Scrap Recycling Industries in the United States, Zorba is defined as a mixture of **shredded** nonferrous scrap metals **primarily** consisting of aluminum but also containing copper, lead, brass, zinc, tin, nickel and copper in any of their forms. Given the fact that we have already seen that using recycled aluminum can result in big savings, and the demand for other nonferrous metals found in Zobra, the commercial potential of this mixture as well as the economic significance of extracting and sorting out these elements from the mixture becomes rather obvious. Typically, these metals would be sorted either manually or by using "sink-float" **gravimetric** treatments.

11 However, both methods are largely unreliable as the first method relies on human **intervention** and observation, while the second method relies on the density of said materials. The problem arises because some of the nonferrous metals are of similar densities so they will not be separated from each other using gravimetric techniques. Using sensors, including for example X-ray transmission technology which can target different materials based on their atomic density, is significantly more reliable and almost completely removes **arbitrariness** enabling the enhanced separation of the scrap materials acquired.

12 For example, in the case of **finely** cut copper containing wires, which are often accompanied by **traces** of brass or stainless steel, using sensor-based technology makes it possible to detect and

remove copper particles smaller than 1mm in size from the mixture ensuring purity of greater than 99%; a rate that could never be matched by most **sophisticated** sink-float mechanisms or the most **scrutinizing** eyes.

13 To sum up, some commonly used methods for recycling and recovering nonferrous metals are sensor-based methods which rely on the use of sensors to detect and sort specific metals, precipitation methods mainly used for recovery from aqueous solutions and wastewater treatment, and electrowinning that has broad application across many different industries.

◆ Words and Expressions

omnipresent / ˌɒmnɪˈprezənt/ *adj.* present everywhere 无所不在的；遍及各处的

submerge / səbˈmɜːdʒ/ *v.* to go under the surface of water or liquid; to put something or make something go under the surface of water or liquid (使)潜入水中；没入水中；浸没；淹没

aqueous / ˈeɪkwɪəs/ *adj.* containing water; like water 水的；含水的；水状的

shred / ʃred/ *v.* to cut or tear something into small pieces 切碎；撕碎

primarily / ˈpraɪmərɪlɪ/ *adj.* mainly 主要地；根本地

intervention / ˌɪntəˈvenʃən/ *n.* action taken to improve or help a situation 出面；介入

arbitrariness / ˈɑːbɪˈtrərɪnɪs/ *n.* the trait of acting unpredictably and more from whim or caprice than from reason or judgment 任意；武断；随心所欲

finely / ˈfaɪnlɪ/ *adj.* into very small grains or pieces 成颗粒；细微的；细小的

trace / treɪs/ *n.* a very small amount of something 微量；少许

sophisticated / səˈfɪstɪkeɪtɪd/ *adj.* clever and complicated in the way that it works or is presented 复杂巧妙的；先进的；精密的

scrutinize / ˈskruːtnaɪz/ *v.* to look at or examine somebody/something carefully 仔细查看；认真检查；细致审查

◆ Terminology

electrowinning / ɪˈlektrəʊˈwɪnɪŋ/ *n.* 电解冶金法；电解沉积

leachate / ˈliːtʃeɪt/ *n.* 沥滤物；沥滤液

anode / ˈænəʊd/ *n.* 阳极；(电解池的)正极；(原电池的)负极

cathode / ˈkæθəʊd/ *n.* 阴极；(电解池的)负极；(原电池的)正极

electrode / ɪˈlektrəʊd/ *n.* 电极

adsorption / æd'sɔːpʃən/ *n.* 吸附(作用)

precipitant / prɪ'sɪpɪtənt/ *n.* 沉淀剂

gravimetric / ˌɡrævɪ'metrɪk/ *adj.* (测定)重量的;重量分析的

☞ Section D Language Focus

Task One | Are the following statements True or False according to the passage? Write T/F accordingly.

1. It is likely, in some situations, to save up to 90% of energy costs if recycling is used as the major production method for aluminum. (   )

2. Waste including nonferrous metals is put into a liquid solution, and then it is changed into liquid state. Such process is called a leachate. (   )

3. Precipitation is also the most widely used method for metal recovery from liquid solutions.
(   )

4. Nonferrous metal sensors are popular only in extracting nonferrous metals from scrap.
(   )

5. It is possible to use sensor-based technology to detect and remove copper particles from finely cut copper containing wires. (   )

Task Two | Translate the following sentences into Chinese.

1. Fortunately, nonferrous metals, especially nickel, silver, copper, aluminum and tin all share the property of being able to undergo unlimited number of instances of recycling without losing any of their original properties.

2. Electrowinning, which is also known as electroextraction, is, on the surface of it at least, a relatively simple process of extracting dissolved metals from their dissolved states using electricity.

3. Precipitation is the process of forming of an insoluble solid material from what originally was an aqueous solution typically involving pH adjustment or addition of another chemical species.

_____

_____

4. According to the Institute for Scrap Recycling Industries in the United States, Zorba is defined as a mixture of shredded nonferrous scrap metals primarily consisting of aluminum but also containing copper, lead, brass, zinc, tin, nickel and copper in any of their forms.

_____

_____

5. However, both methods are largely unreliable as the first method relies on human intervention and observation, while the second method relies on the density of said materials.

_____

_____

Task Three | Choose the best answers for each blank of the following passage.

Zorba, as well as being a famous Greek literary 1._____, is the collective term for shredded and pre-treated nonferrous scrap metals. The Institute of Scrap Recycling Industries (ISRI) in the U.S. 2._____ Zorba as a "shredded mixed nonferrous metals consisting 3._____ of aluminium generated by eddy-current separator or other segregation techniques". Zorba is mainly aluminium but may also 4._____ copper, nickel, stainless steel, tin, zinc, lead and magnesium.

Zorba 5._____ in three size fractions: large, small and fine. The market has evolved to treat each grade 6._____. It prefers the large fraction because it is the 7._____ to sort using sizing equipment and dense media sorting, or by hand sorting in China. It is still possible to sort the small fraction by hand, particularly for the higher value metals in the mix. These can include copper, brass and 8._____ of precious metals.

The fine fraction is not suitable for 9._____ sorting. Dense media separation plants 10._____ sort this fraction. However, they will progressively switch to sensing and sorting

Unit 4 Introduction to Nonferrous Metallurgy

technologies 11. _____ on x-ray transmission (XRT). Sources tell me that two of the three plants of this type in the UK have already made the switch to XRT from dense media separation.

Sorting Zorba into wrought and cast aluminium alloys using XRT is already 12. _____. The rates are high 13. _____ 30 tonnes/hour using conveyor belts that are 2.4 metres wide. A rapidly developing technology is LIBS (Laser Induced Breakdown Spectroscopy). This separates wrought alloys into AA6xxx and AA5xxx families. 14. _____, the highest rates presently available are of the order of 3 tonnes/hour. To date, sorting prompt rather than end-of-life aluminium scrap is the main use for LIBS separation 15. _____.

1. A. term   B. character   C. book   D. writer
2. A. define   B. describe   C. name   D. refer
3. A. partly   B. hardly   C. primarily   D. greatly
4. A. exclude   B. contain   C. include   D. detain
5. A. comes   B. goes   C. keeps   D. gets
6. A. similarly   B. relatively   C. irrelevantly   D. differently
7. A. hardest   B. easiest   C. nicest   D. highest
8. A. lots   B. little   C. traces   D. much
9. A. machine   B. human   C. animal   D. hand
10. A. typically   B. mainly   C. barely   D. mostly
11. A. relied   B. worked   C. based   D. used
12. A. impossible B. reachable   C. unreachable   D. possible
13. A. at   B. in   C. of   D. for
14. A. Thus   B. However   C. And   D. Though
15. A. method   B. skill   C. technology   D. way

Task Four | The following passage has four paragraphs. Choose the correct topic sentence for each paragraph from the list of sentences below.

A. With lockdowns due to the pandemic, governments and the entire metal industry is seeking to complement other metals such as copper for scrap.
B. Development of intelligent sorting systems to help in the sorting and upcycling of nonferrous scrap is creating opportunities in the scrap metal recycling market.
C. Export bans on scrap in some countries has also hit the scrap recycling market.
D. Nonetheless, the scrap metal recycling market is expected to counter headwinds of the COVID-19 pandemic and emerge resilient.
E. The scrap metal recycling market is facing the brunt of the COVID-19 pandemic.

1. _____In developed countries such as the U. K. , furloughs across business sectors as a repercussion of the pandemic has hit the scrap metal recycling sector as well. As a result of which, many scrap yards have shut down or are working at reduced capacities. Besides this, the pandemic has impacted shipment of metals in the country. Due to this, nearly all aluminum smelters in the U. K. have closed to result in price of aluminum to underperform.

2. _____For example, in the UAE and South Africa, scarcity of scrap has led to bans on the export of scrap metal. This has forced some countries to sustain with domestic reserves and increase dependency on other metals for economic activities. Interestingly, this has led to increasing in demand for some other metals such as copper.

3. _____COVID-19 pandemic and emerge resilient. As a result, the scrap metal recycling market is predicted to rise at a robust ~6% CAGR between 2020 and 2030, say analysts at TMR. Expanding at this growth rate, the scrap recycling market is predicted to surpass a valuation of US $516.4 bn by 2030.

4. _____The intelligent sorting systems allow high throughput of feedstock for sorting of metals as well as alloy composition through a combination of AI/ML image processing and sensor fusion technologies.

Unit 4 Introduction to Nonferrous Metallurgy

 Part Ⅲ: Academic Skills

Academic Writing Skills(Ⅳ) *

1. Structuring an Argument and Substantiating Claims or Assertions Through Careful Argument

Argument is a difficult skill to master, developed over time and through practice, and by reading scholarly writing. One of the requirements of higher education is that you read widely and with close attention to the text. Through this reading you will be exposed to a range of books, documents and journals written in different styles. It is important to consider how experienced writers present their work and build an effective argument.

Activity: Constructing an argument

1. Select a passage from a scholarly piece of writing in your own discipline which builds a good argument and examine it to see how the author achieves this.
2. Now focus on your own piece of writing. Freewrite for five minutes in full sentences: what is my argument?
3. Read over what you have written and reduce your argument to just 25 words.
4. Answer the following questions.
a. Does my argument have a clear focus? What exactly am I claiming?
b. Does my argument have sound logic? Is there a clear appeal to reason not emotion?
c. What evidence can I provide in support of my argument? What is the literature saying?
d. Does my argument have a clear, logical structure? Does my argument develop through evidence and analysis? Does it lead to a conclusion?

There is no one technique for developing a good argument but argument has a number of key elements including focus, logic and evidence. Good argument shows an ability to express a critical and objective outlook. However, developing an argument takes time and work. In order to construct a strong and logical argument, Leki notes that it is important to avoid these common flaws:

---

* Taken from *Developing Your Academic Writing Skills: A Handbook* by Marian Fitzmaurice & Ciara O'Farrell.

- Exaggeration and unsubstantiated generalisations;
- Oversimplification of your argument or of the opposing argument;
- Logical flaws;
- Appeals to inappropriate authorities;
- Emotionally charged words;
- Out-of-date facts.

Keep your tone controlled and reasonable and remember that a convincing argument always displays the writer's ability to understand the other side of an argument and to appraise opposing points of view. The skills you are developing include the ability to write in an objective tone, to use relevant sources to support your argument and to provide a logical and systematic analysis.

2. The Language of argumentation

Certain phrases are often used in argumentation and some examples are provided below which you can use to help write your argument.

Words and phrases which can be used in argumentation
- Y argues that...
- Y suggests that...
- Y contends that...
- Y makes a case that...
- X develops the argument further by suggesting that...
- X maintains that...
- X claims that...
- X asserts that...
- In contrast, Y states that...
- X concludes that...

Providing a counter argument
- Despite claims that...
- Some would argue that..., but...
- It has been argued that..., however...
- However...
- While a lot of evidence points to this conclusion... there is another aspect to be considered...
- On the contrary...

- On the other hand...
- Some assert that..., but this underestimates the influence of...

Putting forward your own opinion using the passive voice

- The evidence suggests that...
- It will be argued...
- The paper argues...
- The findings indicate...
- These findings suggest...

Part Ⅳ: Extended Reading and Translation

Translate the following paragraphs into English.

2020年7月22日,国家航天局公布了我国首个独立火星探测任务"天问一号"的火星车详情。国家航天局探月与航天工程中心发布的信息显示,火星车高度有1.85米,重量达到240千克左右,配备6个轮子和4块太阳能板,能在火星上以200米每小时的速度行进。

"天问"来自中国伟大诗人屈原的长诗《天问》。《天问》是屈原对于天空、星辰、自然现象、神话和人世等一切事物现象的发问,体现了他对传统理念的疑惑以及追求真理的精神。"天问"这个名字表达了中华民族对于真理追求的坚韧与执着,体现了对自然和宇宙空间探索的文化传承。

Unit 5 Powder Metallurgy

Part Ⅰ: Text A

Section A Warm-up Questions

Discuss with your partners about the following questions.
1. Do you know anything about powdered metals? Could you give some examples of them?
2. Could you tell us how powder metallurgy touches our life?

Section B Listening Practice

Listen to the audio and fill in the blanks with missing information.

3D Printing: Print My Ride

A mass-market carmaker starts customising vehicles individually. Another milestone has been passed 1. _____, popularly known as 3D printing. Daihatsu, a Japanese manufacturer of small cars and a subsidiary of Toyota, an industry giant, announced on June 20th that it would 2. _____ to customise their vehicles with 3D-printed parts. This 3. _____ the kind of individual tailoring of vehicles hitherto restricted to the luxury limousines and sports cars of the super-rich. The service is available only to buyers of the Daihatsu Copen, a tiny convertible two-seater. 4. _____ from their local dealer can choose one of 15 "effect skins", decorative panels 5. _____

_____ in ten different colours. The buyers can then use a website to tinker with the designs further to create exactly the look they want.

Section C Active Reading

What Is Powder Metal Manufacturing?

1 When the goal for a **part** is the longest possible **lifecycle** and highest performance while operating under high temperatures, in a **corrosive** environment, and with the potential for extreme wear, then powder metal manufacturing can offer a very cost-effective option. Not only are the parts strong, robust, and wear-resistant, but the **advent** of **additive manufacturing** has made it possible to minimize their weight, maximize their durability, and create parts that are simply impossible to manufacture by conventional manufacturing methods.

2 Powder metal (PM) parts are created from a very fine metal powder that is compressed and **sintered** to achieve its final shape. This is quite different from cast parts, which begin their life as a **liquefied** metal, or from machine or forged parts, which start off as stock metal. The powder metal manufacturing process makes it possible to create parts with an extremely complex **geometry**. While such parts might be made through casting or machining, the manufacturing costs increase dramatically as the complexity rises. PM parts, however, are cost effective even when the geometry is complicated and can make parts that are impossible to **fabricate** using any other methods.

3 Another key benefit of powder metal is that it is a net or near net manufacturing process that results in minimal waste material. Rather than subtracting geometry and materials in order to create a part, PM does not waste any metal, making it a much more efficient and environmentally sustainable process. Because PM is so closely allied with computer-aided manufacturing, parts can be created to **simultaneously** optimize weight, strength, **stiffness**, and hardness. This can be vital for applications in industries such as **aerospace** where weight must be minimized.

4 Powder metal products can be found in a wide **array** of industries, including aerospace, automotive, marine, and biomedical. Many everyday products may have been created via PM, such as light bulb **filaments**, automotive engine components, the lining of friction brakes, medical devices, and **lubricant** infiltrated **bearings**. PM can also be used for more exotic purposes, such as heat shields used on spacecraft during re-entry, electrical contacts for extremely high current flows, and gas filters. The parts produced by PM can be used as **prototypes** or as fully functional parts. As better methods of manufacturing and heat treatment of PM parts are developed, the applications will continue to grow and PM parts will become even more commonplace.

Unit 5 Powder Metallurgy

5 Powder metallurgy (which forms the basis of modern PM methods and technology) actually dates to the 1940s. Early products made by these methods include **porous** bearings, electrical contacts, and **cemented carbides**. Through the intervening years, companies wisely invested in powder metallurgy technology and advances, focusing their research on aspects such as refinement, new alloy development, and **atomization** techniques for efficiently generating fine powders. Such research and innovation continue to this day. One of the most **groundbreaking** developments is net shape production of PM parts via additive manufacturing (AM).

6 The most commonly used base metals for PM processes include alloyed metals such as:
- Iron
- Steel
- Copper
- Stainless steel
- Titanium
- Aluminum
- Tin
- **Molybdenum**
- Tungsten
- **Tungsten carbide**
- Various precious metals

7 Most industrial PM products are comprised primarily of iron and steel along with other elements, including both metal, semi-metal, and transitional elements. Different alloying elements can be added to the **base metals** to achieve customized or improved material properties.

8 There are many other PM processes that have been successfully developed since the 1940s. The more traditional processes include **powder forging**, **hot isostatic pressing** (HIP), metal injection molding, and electric current assisted sintering.

9 In powder forging, a pre-form is made using the conventional press and sinter methods, but the part is then heated and hot forged. The result is a full-**density** part with as-wrought properties.

10 In HIP, the powder is gas atomized and spherical in shape. The mold used is typically a metal can of appropriate shape into which the powder is added. The mold is sealed, vibrated, and then all air is vacuumed out via a pump. The mold is then placed into a hot isostatic press where it is heated to a **homologous** temperature and its internal pressure is increased via external gas pressure. This particular PM process results in a finished part that has the correct shape and full density. The part has as-wrought or better mechanical properties.

11 Hot isostatic pressing was developed in the late 1950s and early 1960s and entered tonnage

production in the 1970s. In 2015, HIP was used to manufacture approximately 25000 tons of stainless steel and tools steels annually. This, in turn, led to the manufacture of super alloys used in aerospace and jet engines.

12   Another common manufacturing technique for PMs is **metal injection molding** (MIM). During this process, spherical metal powder that is less than 25 **microns** in size is mixed with either plastic or wax as a **binding agent**. Once a near solid part has been formed (65% volume), it is injection-molded. The result is a "green" part that typically has a very complex geometry. The green part is then heated under controlled conditions to remove the binder in a process known as de-bindering. The part at this stage is referred to as a "brown" part, but the process is still not complete. The brown part is then subject to an atmospherically controlled sintering process. The part's volume is reduced by about 18%, and the final part is extremely dense at 97%~99%.

13   **Electric current assisted sintering** (ECAS) is a different type of powder metal manufacturing process that makes extensive use of electric currents and does not require the use of binders. Instead of de-binding or sintering after pressing, electric currents are used to increase the density of the powder, which significantly reduces the thermal cycle needed to maintain the strength and density of the final part. This in turn reduces the overall production time for the part. For example, some parts will see a process time reduction from 15 minutes down to just a few **microseconds**. However, this process only works on relatively simple shapes. Another interesting aspect of the ECAS process is that the molds used are actually designed for the final part shape because the powders achieve final density while filling the mold under pressure and heat. This takes care of both **distortion** and shape variation.

14   Additive manufacturing (AM), sometimes called metal 3D printing, is considered a newer PM method, although its history dates back to the 1980s. In this method, parts are formed by melting or laser sintering metal powders (as well as other forms of metals, ceramics, and **polymers**) and additives. What makes these two methods so different from the other PM methods is the layer-by-layer approach taken to build up the part (hence the term additive manufacturing, which refers to the process of adding a layer at a time to form the part). This allows parts to be built a single layer (micrometers thick) at a time based directly on 3D digital models of the part via computer aided manufacturing (CAM). This layered approach is digitally controlled to achieve a high level of precision.

15   AM supports the manufacturing of very **intricate**, complex geometries that are often impossible to create using other methods such as metal casting or machining. AM is known for being a highly customizable, versatile, flexible design process that supports not just metals but hybrids, composites, and even **functionally graded materials** (FGM). Materials that can be used include

hybrids and composites, metals, polymers, nanomaterials, **pharmaceuticals**, biological materials, and ceramics.

◇ Words and Expressions

part / pɑːt/ *n.* a piece of a machine or structure 部件；零件

lifecycle / ˈlaɪfˌsaɪkl/ *n.* the period of time during which something, for example a product, is developed and used 生命周期；寿命（产品等从开发到使用完毕的一段时间）

corrosive / kəˈrəʊsɪv/ *adj.* tending to destroy something slowly by chemical action 腐蚀性的；侵蚀性的

advent / ˈædvənt/ *n.* the coming of an important event, person, invention, etc. （重要事件、人物、发明等的）出现；到来

fabricate / ˈfæbrɪkeɪt/ *v.* to make or produce goods, equipment, etc. from various different materials 制造；装配；组装

simultaneously / sɪmlˈteɪniəsli/ *adv.* at the same instant 同时发生（或进行）地；同步地

stiffness / ˈstɪfnɪs/ *n.* the physical property of being inflexible and hard to bend 不易弯曲；坚硬

aerospace / ˈeərəspeɪs/ *n.* the industry of building aircraft, vehicles and equipment to be sent into space 航空航天工业

array / əˈreɪ/ *n.* a group or collection of things or people, often one that is large or impressive 大堆；大群；大量

filament / ˈfɪləmənt/ *n.* a thin wire in a light bulb that produces light when electricity is passed through it （电灯泡的）灯丝；丝极

lubricant / ˈluːbrɪkənt/ *n.* a substance, for example oil, that you put on surfaces or parts of a machine so that they move easily and smoothly 润滑剂；润滑油

prototype / ˈprəʊtətaɪp/ *n.* the first design of something from which other forms are copied or developed 原型；雏形；最初形态；样品；模型

porous / ˈpɔːrəs/ *adj.* having many small holes that allow water or air to pass through slowly 多孔的；透水的；透气的

groundbreaking / ˈgraʊndbreɪkɪŋ/ *adj.* making new discoveries; using new methods 开创性的；创新的；革新的

homologous / hɒˈmɒləgəs/ *adj.* similar in position, structure, etc. to something else （位置、结构等）相应的；类似的；同源的

distortion / dɪˈstɔːʃn/ *n.* change in shape 变形

intricate / ˈɪntrɪkət/ *adj.* having a lot of different parts and small details that fit together 错综复杂的

pharmaceutical / ˌfɑːməˈsuːtɪkl/ *n.* a drug or medicine 药物

◇ Terminology

sinter / ˈsɪntə/ *v.* (使)烧结；(使)熔结

liquefy / ˈlɪkwɪfaɪ/ *v.* (使)液化

geometry / dʒɪˈɒmɪtrɪ/ *n.* 几何形状；几何图形；几何结构

bearing / ˈbeərɪŋ/ *n.* (机器的)支座；(尤指)轴承

atomization / ˌætəumaiˈzeɪʃən/ *n.* 雾化；[分化]原子化

molybdenum / məˈlɪbdənəm/ *n.* 钼

density / ˈdensɪtɪ/ *n.* 密度(固体、液体或气体单位体积的质量)

micron / ˈmaɪkrɒn/ *n.* 微米

microsecond / ˈmaɪkrəʊˌsekənd/ *n.* 微秒

polymer / ˈpɒlɪmə(r)/ *n.* 聚合物；多聚体

additive manufacturing 增材制造；添加剂制造

cemented carbide [材]硬质合金；烧结硬质合金

tungsten carbide 硬质合金；[无化]碳化钨

base metal 贱金属；基体金属

powder forging [机]粉末锻造

hot isostatic pressing 热等静压；高温等静力压制

metal injection molding 金属粉末注射成型

binding agent [胶粘]粘合剂；黏合剂

electric current assisted sintering 电流辅助烧结

functionally graded materials 功能梯度材料

Unit 5 Powder Metallurgy

Section D Language Focus

Task One | Text organization

Work in groups and discuss the organization of the text and fill in the blanks.

Parts	Paragraphs	Main Ideas
Part 1	Para. 1	Brief introduction to 1._____
Part 2	Para. 2-3	Powder metals and their 2._____
Part 3	Para. 4-7	3._____ on power metallurgy and 4._____ used
Part 4	Para. 8-13	Traditional PM 5._____
Part 5	Para. 14-15	Introduction to 6._____

Task Two | Answer the following questions based on the information contained in the text.

1. Which statement about powder metals is TRUE? _____

   A. Powder metals start off as a liquefied metal.

   B. Powder metals begin their life as stock metal.

   C. Powder metals are made from a very fine metal powder.

   D. Powder metals can be hardly created to optimize weight.

2. What is NOT the benefit of powder metal? _____

   A. Minimizing the weight.

   B. Environmental sustain.

   C. Waking good use of the metal.

   D. Saving money.

3. What is one of the greatest developments of PM parts via additive manufacturing? _____

   A. Refinement.

   B. Net shape production.

   C. Alloy development.

   D. Atomization techniques.

4. Which PM process is helpful for producing super alloys used in aerospace and jet engines? _____

   A. Powder forging.

   B. Hot isostatic pressing (HIP).

C. Metal injection molding.

D. Electric current assisted sintering.

5. What approach makes additive manufacturing different from most other PM methods? _____

A. Layered approach.

B. Electric currents approach.

C. Injection approach.

D. Pressing approach.

Task Three | Fill in the blanks with the words or phrases given in the box. You may not use any of the words or phrases more than once.

The global additive manufacturing with metal powders market is expected to reach USD 50.65 billion by 2026, 1._____ to a new report by Reports and Data. This can be mainly 2._____ with increasing demand in the automotive industry globally. Based on statistics, Affordable Manufacturing cost is expected to become the most common growth 3._____ globally in the coming years. High customer requirements and global enhanced demand from aerospace and automation industry are also significant 4._____ stimulating market demand.

The term Additive Manufacturing is also known as the 5._____ of objects through the deposition of a material by using a print head, nozzle, along with another printer technology. Additive manufacturing with metal powders is the process of 6._____ elements to make objects from 3D model data, usually 7._____ upon layer, as exposed to subtractive production methodologies such as machining.

North America is expected to be a key 8._____ generating region in the forecast period. The market is projected to grow at a CAGR of 14% in the forecast period. Increasing growth of technologies in the region has 9._____ a growth opportunity for the increasing growth of the market. Manufacturers have been focusing on 10._____ out new technologies.

Unit 5　Powder Metallurgy

A. interest	I. revenue
B. related	J. created
C. fabrication	K. working
D. according	L. due
E. bringing	M. layer
F. associated	N. factors
G. methods	O. condition
H. joining	

Task Four │ The following passage has four paragraphs. Choose the correct topic sentence for each paragraph from the list of sentences below.

A. The industry is turning to high grade, hard rock lithium pegmatite deposits.
B. We're facing an imminent crisis in global lithium markets.
C. Global lithium supply is concentrated in relatively few locations.
D. Lithium is in the midst of an unprecedented boom.
E. Traditional lithium brine projects take too long to put into production.

1. _____ Demand is growing exponentially, and lithium consumers are facing a 100,000-ton shortfall by 2025, according to oilprice.com.

2. _____ At up to 48 months, they won't be able to close the supply gap on their own. Mentioned in today's oilprice.com commentary include: Pengrowth Energy Corp., Franco-Nevada Corporation, Cameco Corporation, Ballard Power Systems, Hydrogenics.

3. _____ That's why Power Metals Corp is becoming a critical stock to follow. They've recruited the world's top lithium pegmatite expert and invested in next generation 3D modelling for what could be the largest drill campaign of its kind for lithium today.

4. _____ Since 2015, the price per ton has soared from $6500 to over $20000, and demand shows no signs of stopping. In fact—with explosive demand growth for smartphones, EV's and home storage batteries—we might soon hit the physical limits of our lithium supply chain. Companies like Power Metals stand to attract major investor attention.

Task Five | Translate the following sentences into English, using the words or phrases in the brackets.

1. 在三维制造的经典场景里，我们先在电脑的建模软件里创建一个三维模型。(fabrication)

2. 可以使用润滑剂，比如硅系列油，使耗尽了油的滚珠轴承的受损表面更加光滑，避免过快出现划痕。(lubricant, ball-bearings)

3. 我们已经3D打印了所有原型，这样就可以快速建立设计网络。(prototype)

4. 这些是多孔的分子网络，空气可以通过它们循环。它们的多孔性使它们具有很大的表面积。(porous)

5. 一个零件的收缩率高达35%，而且用于其他工艺的简单收缩模型无法预测成型后烧结过程中的变形。(distortion)

Unit 5 Powder Metallurgy

Part Ⅱ: Text B

Section A Warm-up Questions

Discuss with your partners about the following questions.
1. Do you think copper is a green metal? Why or why not?
2. Do you know properties of copper?

Section B Listening Practice

Listen to the audio clips and fill in the blanks with missing information.

There is only one metal which is 1. _____. Its name is copper. On the shores of Lake Superior in North America, lumps of pure copper are found in the rocks; but in Cornwall, and in Australia, and in many other places, we find copper as an ore.

An ore is a metal united with 2. _____. Copper ore looks like a stone of a black, or a green, or a red colour. Looking at a lump of the ore, no one would think that there was any copper in it.

The other substances which are combined with the copper to form the stony-looking ore are 3. _____. These substances are got rid of by first roasting the ore in a furnace, and then smelting or melting it in a much hotter furnace. This is repeated three times, until at last the pure copper is left by itself.

The 4. _____ is chiefly carried on in the town of Swansea, in South Wales, to which port ships bring copper ore from all parts of the world.

Copper can be rolled or hammered out into thin sheets, and so we say that it is malleable. It can also be easily drawn out into wire, and for this reason 5. _____. Sheet copper is much used for covering the bottoms of ships, and for making kettles and saucepans and many other household articles.

Section C Active Reading

How Hydrometallurgy and the SX/EW Process Made Copper the "Green" Metal

1 Copper is traditionally known as the "red" metal after its natural color. However, it is also known as a "green" metal for the green **patina** that it acquires due to **weathering**. Indeed, **patinized** copper is the architectural **focal** point of many modern buildings for its natural look. Beyond this, however, copper can truly be cited as the "green" metal both for its role in protecting the natural environment through its use in energy-saving applications and for the achievements that have been realized in the production of the metal in an environmentally sound manner.

2 The energy efficiency resulting from the use of copper in high efficiency motors, electrical transformers, underground power lines, air conditioning and refrigeration equipment, electric vehicles, etc. has a significant impact on the release of green house gases resulting from the generation and use of fossil fuel based electrical power. Likewise, newly developed, high efficiency automobile **radiators** reduce fuel consumption by being smaller, lighter and having a lower pressure drop than their aluminum **counterparts**.

3 In both cases, the savings in energy consumed also means a conservation of our fossil fuel resources and the reduction of greenhouse gases.

4 The production of copper, as in the utilization of any other natural resource, has an impact on the environment. This cannot be avoided since the earth must be disturbed in order to extract copper from it; however, the object of the copper mining industry has been to make this impact as small as possible. Significant improvements have been made in environmental impact as new technologies have been applied to the production of copper. Great **strides** have been accomplished in the conventional treatment of copper ores, such as at the Bingham Canyon mines, in Utah and the **adjoining** copper smelters.

5 Conventionally, copper is recovered by a **pyrometallurgical** process known as smelting. In this process copper ore is mined, crushed, ground, concentrated, smelted and refined. The mining, crushing and grinding portions of the processing are extremely energy intensive since the rock must be reduced essentially to **talcum** powder fineness in order to separate the copper-bearing minerals from it. To be applicable to this process, the ores must contain copper minerals in sulfide form; as mineral such as **Chalcocite** (Cu_2S), **Chalcopyrite** ($CuFeS_2$) and **Covellite** (CuS). In the concentrating operations these minerals are separated from the gangue material of the ore, that might contain as little as 0.5% copper to form a **concentrate** containing 27% to 36% copper. In the

smelting operation, the concentrate is **fed** to a **smelter** together with oxygen and the copper and iron sulfides are oxidized at high temperature resulting in impure molten metallic copper (97% to 99%), molten iron oxide and gaseous sulfur dioxide. The impure copper is then purified by electrolytic purification to 99.99% pure copper while the iron oxide is **disposed of** as slag.

6 Typically, in this process there is more sulfur dioxide produced by weight than there is copper. Rather than discharge the sulfur dioxide into the air, as was once the practice, the sulfur dioxide is captured and converted into **sulfuric acid**. In the United States, smelter-produced sulfuric acid amounts to approximately 10% of total acid production from all sources. Prior to the mid-1980s, this by-product sulfuric acid had to be sold to other industries, often at a loss due to the long shipping distances.

7 Beginning in the mid-1980s a new technology, commonly known as the **leach-solvent extraction**-electrowinning process or, SX/EW Process, was widely adopted. This new copper technology utilizes smelter acid to produce copper from oxidized ores and mine wastes. Today, worldwide, approximately 20% of all copper produced is produced by this process. In Latin America, the total is closer to 40% whereas in the United States the total is approaching 30%.

8 The SX/EW Process is a **hydrometallurgical** process since it operates at **ambient temperatures** and the copper is in either an aqueous environment or an organic environment during its processing until it is reduced to the metal. Because of its dependence on sulfuric acid, the SX/EW Process is at present not a substitute for, but rather an adjunct to conventional smelting. However, it is also applicable in locations where smelter acid is not available by the purchase of sulfuric acid or the manufacture of sulfuric acid from sulfur or **pyrite**. In addition, it offers the opportunity to recover copper from an entirely different set of ores and mining byproducts than is possible by smelting; namely, oxidized materials. These may be mined copper minerals that are in oxidized form—minerals such as **Azurite** ($2CuCO_3 \cdot Cu(OH)_3$), **Brochantite** ($CuSO_4$), **Chrysocolla** ($CuSiO_3 \cdot 2H_2O$) and **Cuprite** (Cu_2O), residual copper in old mine waste dumps whose sulfide minerals have been oxidized by exposure to the air or sulfidic copper minerals that have been oxidized by another new technology—bacterial leaching. In addition, the process can be used to extract copper **in situ**. That is, without removing the material from the waste pile or from the ground. The net result of the use of this process is that copper can be produced from sources that in the past would have gone untouched, thus reducing the reliance on conventional ore bodies. Further, the process is capable of removing copper from waste materials where otherwise it would have been considered a **contaminant** to the environment. In the United States, for example, copper is considered to be a toxic material released to the environment once it is mined under Emergency Planning and Community Right-to-Know Act (EPCRA) and the Environmental Protection Agency's

Toxic Release **Inventory** (TRI). Copper mine dumps and flotation **tailings** constitute a significant inventory of copper that is considered to be a contaminant to the environment under TRI.

9 The SX/EW process, itself, has very little environmental impact because its liquid streams are very easily contained. There is no **effluent inasmuch as** all impurities are returned to the site where they originated and the sulfuric acid is eventually neutralized by the limestone in the ore body or waste dump where it is deposited as calcium sulfate (**gypsum**)—a very insoluble substance.

10 The process involves leaching the material with a weak acid solution. This solution, known as **pregnant liquor**, is recovered and then contacted with an organic solvent, referred to as the extractant, in the solvent extraction stage (SX). Here the copper is extracted away from the aqueous phase leaving behind most of the impurities that were in the leach solution. Since the copper ion is exchanged for hydrogen ion, the aqueous phase is returned to its original acidity and recycled to the leaching step of the process. Meanwhile, the copper-bearing organic phase is **stripped** of its copper by contacting it with a strongly **acidified** aqueous solution at which time the copper is moved to the aqueous phase while the organic phase is reconstituted in its hydrogen form. The copper-bearing aqueous phase is advanced to the electrowinning (EW) stage of the process while the barren organic phase is returned to the extraction stage of the process. In the electrowinning stage of the process the copper is reduced electrochemically from copper sulfate in solution to a metallic copper cathode. Electrowon copper cathodes are as pure as or purer than **electrorefined** cathodes from the smelting process. Thus they are well received by the market.

◆ Words and Expressions

focal / ˈfəʊkl/ *adj.* central; very important; connected with or providing a focus 中心的；很重要的；焦点的；有焦点的

counterpart / ˈkaʊntəpɑːt/ *n.* a person or thing that has the same position or function as somebody/something else in a different place or situation 职位(或作用)相当的人；对应的事物

stride / straɪd/ *n.* an improvement in the way something is developing 进展；进步；发展

adjoin / əˈdʒɔɪn/ *v.* to be next to or joined to something 紧挨；邻接；毗连

conventional / kənˈvenʃənl/ *adj.* following what is traditional or the way something has been done for a long time 传统的；习惯的

feed / fiːd/ *v.* to put or push something into or through a machine 把……放进机器；将……塞进机器

contaminant / kənˈtæmɪnənt/ *n.* a substance that makes something impure 致污物；污染物

inventory / ˈɪnvəntrɪ/ *n.* a written list of all the objects, furniture, etc. in a particular

Unit 5 Powder Metallurgy

building(建筑物里的物品、家具等的)清单;财产清单

effluent / ˈefluənt/ *n*. liquid waste, especially chemicals produced by factories, or sewage 流出物,流出液(尤指工厂排出的化学废料)

strip / strip/ *v*. to remove a layer from something, especially so that it is completely exposed 除去;剥去(一层)

dispose of to get rid of somebody/something that you do not want or cannot keep 去掉;清除;销毁

in situ 在原位;在原地;在合适地方

inasmuch as *conj*. 因为;由于

◆ Terminology

patina / ˈpætɪnə/ *n*. (金属表面的)绿锈,铜锈,氧化层

weathering / ˈweðərɪŋ/ *n*. (岩石的)风化

patinize / ˈpætɪˌnaɪz/ *v*. 生绿锈

radiator / ˈreɪdɪeɪtə/ *n*. (车辆或飞行发动机的)冷却器,水箱

pyrometallurgical / paɪrɒmɪtæˈlɜːdʒɪkəl/ *adj*. 火法冶金的

talcum / ˈtælkəm/ *n*. [矿物] 滑石

chalcocite / ˈkælkəsaɪt/ *n*. [矿物] 辉铜矿

chalcopyrite / ˌkælkəˈpaɪraɪt/ *n*. [矿物] 黄铜矿

covellite / kəʊˈvelaɪt/ *n*. [矿物] 铜蓝;靛铜矿

concentrate / ˈkɒnsntreɪt/ *n*. 浓缩物

smelter / ˈsmeltə/ *n*. 熔炉

hydrometallurgical / ˈhaɪdrəʊˌmetəˈlɜːdʒɪkəl/ *adj*. 湿法冶金学的;水冶的

pyrite / ˈpaɪraɪt/ *n*. [矿物] 黄铁矿

azurite / ˈæʒʊraɪt/ *n*. 蓝铜矿

brochantite / ˌbrɔtʃænˈtaɪt/ *n*. [矿物] 水胆矾

chrysocolla / krɪsəˌkɔlə/ *n*. [矿物] 硅孔雀石

cuprite / ˈkjuːpraɪt/ *n*. [矿物] 赤铜矿

tailing / ˈteɪlɪŋ/ *n*. 残渣;尾料

gypsum / ˈdʒɪpsəm/ *n*. 石膏

acidify / əˈsɪdɪfaɪ/ *v*. (使)变成酸,酸化

electrorefine / elektrɔːrɪˈfaɪn/ *v*. 电解提纯;电精炼

sulfuric acid *n.* 硫酸

leach-solvent extraction 浸出溶剂萃取

ambient temperature 常温

pregnant liquor 母液

Section D Language Focus

Task One | Are the following statements True or False according to the passage? Write T/F accordingly.

1. Copper is known as a "red" and "green" metal.　　　　　　　　　(　　)

2. The production of copper has nearly no impact on the environment.　(　　)

3. Smelting method for copper production is a kind of energy-saving way.　(　　)

4. The SX/EW Process nowadays does not replace conventional copper extraction owing to its reliance on sulfuric acid.　(　　)

5. The SX/EW Process involves leaching the material with a strong acid solution, known as pregnant liquor.　(　　)

Task Two | Translate the following sentences into Chinese.

1. Likewise, newly developed, high efficiency automobile radiators reduce fuel consumption by being smaller, lighter and having a lower pressure drop than their aluminum counterparts.

2. In the smelting operation, the concentrate is fed to a smelter together with oxygen and the copper and iron sulfides are oxidized at high temperature resulting in impure molten metallic copper (97% to 99%), molten iron oxide and gaseous sulfur dioxide.

Unit 5 Powder Metallurgy

3. In the United States, smelter-produced sulfuric acid amounts to approximately 10% of total acid production from all sources. Prior to the mid-1980s, this by-product sulfuric acid had to be sold to other industries, often at a loss due to the long shipping distances.

_____

_____

4. The net result of the use of this process is that copper can be produced from sources that in the past would have gone untouched, thus reducing the reliance on conventional ore bodies.

_____

_____

5. Meanwhile, the copper-bearing organic phase is stripped of its copper by contacting it with a strongly acidified aqueous solution at which time the copper is moved to the aqueous phase while the organic phase is reconstituted in its hydrogen form.

_____

_____

Task Three | Choose the best answers for each blank of the following passage.

A technology that has aided 1. _____ the production of environmentally clean copper is that of bacterial leaching or bioleaching. This is used as an adjunct to the SX/EW process in situations 2. _____ sulfide copper minerals must be leached. Modern commercial application of bacterial leaching 3. _____ in the 1950s at Kennecott's Bingham mine near Salt Lake City, Utah. It was noticed that blue copper-containing solutions were 4. _____ out of waste piles that contained copper sulfide minerals—a condition that should not have happened in the 5. _____ of powerful oxidizing agents and acid. On investigation it was found that naturally occurring bacteria were oxidizing iron sulfides and the resulting ferric sulfate was acting 6. _____ an oxidizer and leachant for copper sulfides. These bacteria were given the name—ferrooxidans for their action in oxidizing iron sulfides. A 7. _____ set of bacteria were also identified and given the name—thiooxidans for their action in oxidizing sulfur to yield sulfuric acid.

Bacterial leaching, combined with SX/EW, offers a method of exploiting 8._____ ore bodies with a minimum of capital investment. Most commercial operations leaching copper 9._____ ore dumps are located in the Southern Hemisphere in Australia, Chile, Myanmar and Peru. The process 10._____ of injecting the material to be leached with cultivated strains of appropriate bacteria and maintaining conditions that are 11._____ to their effective operation and propagation. Air, for instance, is 12._____ into the heap through air lines situated under the leach pad. Since these naturally-occurring bacteria are present nearly everywhere, they 13._____ play a role in all acid leaching operations; however, these are not considered bacterial leaching since neither cultivated bacteria nor air are added and acid is applied. 14._____ copper bacterial leaching thus far has been confined to the leaching of ore, pilot plant tests are underway for the leaching of chalcopyrite concentrates that would normally be 15._____ by smelting.

1. A. with   B. in   C. for   D. from
2. A. where   B. which   C. that   D. when
3. A. begins   B. invents   C. began   D. invented
4. A. going   B. gone   C. run   D. running
5. A. absence   B. presence   C. absent   D. present
6. A. for   B. with   C. as   D. in
7. A. first   B. second   C. other   D. another
8. A. small   B. big   C. part   D. half
9. A. of   B. with   C. for   D. from
10. A. includes   B. consists   C. deprives   D. strips
11. A. conductive B. helpful   C. assistant   D. harmful
12. A. driven   B. flown   C. blown   D. floated
13. A. unintendedly B. undoubtedly C. unthankfully   D. unluckily
14. A. Because   B. Since   C. While   D. For
15. A. processed   B. accessed   C. addressed   D. projected

Unit 5   Powder Metallurgy

Task Four | The following passage has four paragraphs. Choose the correct topic sentence for each paragraph from the list of sentences below.

A. For non-electrical purposes, copper is also used to make large quantities of plumbing tube, roofing sheet and heat exchangers.
B. The usual commercial supplies of pure copper are used for the most critical of electrical applications such as the production of fine and superfine enamelled wires.
C. Good quality high conductivity copper can be recycled by simple melting and check analysis before casting, either to finished shape or for subsequent fabrication.
D. Where scrap copper is associated with other materials, for example after having been tinned or soldered, it will frequently be more economic to take advantage of such contamination than try to remove it by refining.
E. The copper used for power cables is also drawn from high conductivity rod but to a thicker size than fine wires.

1. _____ It is essential that purity is reproducibly maintained in order to ensure high conductivity, consistent anneal ability and freedom from breaks during rod production and subsequent wire drawing. Since the applied enamel layers are thin but have to withstand voltage, they must have no surface flaws; consequently the basis copper wire must have an excellent surface quality. Primary copper of the best grade is used for producing the rod for this work. Uncontaminated recycled process scrap and other scrap that has been electrolytically refined back to grade "A" quality may also be used.

2. _____ The quality requirements are therefore slightly less stringent. The presence of any undesirable impurities can cause problems such as hot shortness which gives expensive failures during casting and hot rolling. For the same reason, scrap containing such impurities can only be used for this purpose if well diluted with good quality copper.

3 _____ High electrical conductivity is not mandatory and other quality requirements are not so onerous. Secondary copper can be used for the manufacture of these materials, though still within stipulated quality limits for impurities.

4. _____ Many specifications for gunmetals and bronzes require the presence of both tin and lead so this type of scrap is ideal feedstock. Normally it is remelted and cast to ingot of certified analysis before use in a foundry. Scrap of this type commands a lower price than uncontaminated copper.

Part III: Academic Skills

Academic Writing Skills (V) *

1. Finding Your Voice

Voice is a difficult concept to define but developing a voice is an important aspect of becoming an effective writer. Every piece of writing has a voice; voice refers to the way we reveal ourselves to others when we write. Voice may be thought of as "a combination of the personality of the writer that comes through to the reader; the perspective the writer assumes, often influenced by the audience being addressed, as well as by the purpose and previous levels of knowledge, [...] and the tone of the passage". However for a novice writer, it can be difficult to find and express your voice.

Remember that each one of us approaches a topic from a different perspective, so we can all make a valuable contribution to debate. Your distinctive individual perspective might reflect your life experiences, your educational background or your philosophical values. There is no expert or scholar whose answers are so perfect that the rest of us need no longer give our opinions on the topic.

Activity: Analysing a passage for voice

Read the passage below and answer the questions that follow.

> The term "narrative" carries many meanings and is used in a variety of ways by different disciplines, often synonymously with "story". I caution readers not to expect a simple, clear definition of narrative here that can cover all applications, but I will review some definitions in use and outline what I think are the essential ingredients. Briefly, in everyday oral storytelling, a speaker connects events into a sequence that is consequential for later action and for the meanings that the speaker wants listeners to take away from the story. Events perceived by the speaker as important are selected, organized, connected, and evaluated as meaningful for a particular audience. Later chapters will expand and complicate the simple definition with research based on spoken, written, and visual materials... The concept of narrative has achieved a degree of popularity that few would have predicted when some of us, several decades ago, began working with stories that developed in research interviews and medical consultations. More than ten years ago, I began to be uneasy about what I called the tyranny of narrative, and the concern has only increased. It is not appropriate to police language, but specificity has been lost with popularisation. All talk and text is not narrative.

---

* Taken from *Developing Your Academic Writing Skills: A Handbook* by Marian Fitzmaurice & Ciara O'Farrell.

Unit 5 Powder Metallurgy

> 1. Analyse the passage for features of voice. Think about whether it sounds informed, authoritative, questioning. Does it sound credible?
> 2. Describe the voice, drawing on specific items in the passage to support your views.

It takes confidence, practice, time and regular writing to develop your voice but the more you write the more you will develop your own voice. Often the overuse of quotes or paraphrasing without any attempt to discuss the points being made shows that the author is having difficulty finding his or her voice. In academic writing, the use of voice is not about emotion or personal experience, but about being clear, concise, accurate and backing up what is being said with evidence, making a judgement and assessing the contribution of other writers.

2. When to Use "I" in Your Writing

In the above piece, the first-person use is evident as it positions the writer in the text. However, in academic writing use of the first person depends on the discipline. Often, the third person and the passive voice will be used to convey your ideas and arguments. However, the first person can be used effectively, especially in introductions and conclusions. Read some articles from your discipline and see what is typical and what works well.

Whenever you write, a certain voice is being revealed. The challenge is to begin to develop your academic voice. So to develop your voice:

- Read widely and critically;
- Note specialised vocabulary;
- Observe textual features (these vary by discipline) including conventions such as style preferences for writing in first or third person;
- Do not overuse the first person;
- Let your ideas flow freely at the start and then impose a structure;
- Remember that your personal voice should be informed and knowledgeable;
- Establish your credibility by ensuring that you have read enough to have substance for your arguments;
- Always respect other perspectives and develop your counterarguments in a respectful tone;
- Avoid broad, sweeping generalisations;
- Try to get feedback on your work. Ask for honest feedback from a colleague or friend.

Part Ⅳ: Extended Reading and Translation

Translate the following passage into English.

共建"一带一路"倡议以共商共建共享为原则,以和平合作、开放包容、互学互鉴、互利共赢的丝绸之路精神为指引,以政策沟通、设施联通、贸易畅通、资金融通、民心相通为重点,已经从理念转化为行动,从愿景转化为现实,从倡议转化为全球广受欢迎的公共产品。

有色行业是国务院明确的"制造能力强、技术水平高、国际竞争优势明显、国际市场有需求"的12个开展国际产能合作的重点行业之一。在积极响应国家"一带一路"倡议方面,有色行业既是重要的参与者,也将成为重要的受益者。参与"一带一路"建设,有利于我国有色行业融入全球资源配置、提升资源保障度。我国矿产资源领域的合作应该作为有色行业企业参与"一带一路"建设的重要产业和先行产业。

Unit 6   Development of Nonferrous Metallurgy

Part Ⅰ: Text A

Section A   Warm-up Questions

Discuss with your partners about the following questions.

1. Do you like robots? Could you share your experience with robots?
2. Could you list the things that can be done by artificial intelligence at present?
3. Do you think robots could replace us at work? Why or why not?

Section B   Listening Practice

Watch the video clips and fill in the blanks with missing words or phrases.

What is AI? Let's start with the question: what do you think the most complex object in the universe is? Try and think of it. I guarantee you the answer is 1. _____. That's because it's the human brain. The most complex networks, the most powerful systems, cannot match it. 2. _____ that is the ultimate goal of artificial intelligence. It is not about building a robot but creating a computer mind that can think 3. _____ a human, but there are many steps along the way. So-called simple or narrow AI systems are already everywhere. From Apple's Siri to Facebook's 4. _____, it's in our cars, our homes, and air traffic control. And narrow AI has been around for years, doing one 5. _____ task better than any human. The computer Deep Blue 6. _____ the world chess champion way back in 1997, but ask it to play 7. _____ and it wouldn't know where to start. It couldn't learn a new game for itself. It couldn't think as a human. And so we come back to the challenge. Some say the danger of creating a human or general AI, a computer mind that thinks like a human, that learns, that improves, that could even become 8. _____. Experts predict 2050 is the year we could see it, if it is even possible. It's a risk worth billions. Some say it will save 9. _____. Other say it could destroy us. Either way, if and when it happens, the world will be changed 10. _____.

☞ Section C Active Reading

Smart Nonferrous Metallurgy

1 Digitization has become one of the tools of the fourth industrial revolution. Digital technologies are **dramatically** transforming many spheres of our life—retail, banking, communication between citizens and the state. Digitization has also affected industry, especially metallurgy, where both the main and **auxiliary** productions functions are becoming digitized. According to forecasts by Technavio, the industrial AI market will grow at an **accelerated** rate—averaging 54% **per annum**—over the next 3 years. Most major enterprises have already made attempts at in-depth analysis of their accumulated data and creation of models based on artificial intelligence (AI), creating **subdivision**s to handle data analysis, **monetization** and digitization. Nevertheless, AI-based solutions are being **implemented slightly** more slowly in ferrous metallurgy than in other branches.

2 Why is this so and what are the **prospects** for application of AI in nonferrous metallurgy?

3 Given the low **content** of the required component in ore and the **admixtures** of other elements, nonferrous metallurgy is an energy-consuming industry with a complex structure. Thus, ores contain a maximum of 5% copper and 5.5% of **zinc** and **lead**. **Feedstock** is often multicomponent, potentially containing 30 or more chemical elements. To produce one ton of copper, 100 t of ore must be extracted. Energy accounts for half of the total costs related to smelting, with up to 5 kW of energy per hour required to produce 1 t of copper. Many branches of nonferrous metallurgy are also characterized by multi-stage metallurgical conversion of **intermediate** products.

4 All these, on the one hand, are **favourable** factors for the application of AI technologies in this field, since they can often help to increase unit efficiency. On the other hand, nonferrous metallurgy produces quite a large range of fundamentally different products, and, accordingly, uses a very wide **spectrum** of production processes. AI models must **take into account** the specific characteristics of each enterprise when even those working with the same type of metal often differ greatly. All this reduces the potential for **upscaling** any newly created solution and thus reduces the attractiveness of

Unit 6 Development of Nonferrous Metallurgy

investment in research while increasing the implementation risks. As a result, implementation of AI-based solutions in nonferrous metallurgy is more complex and therefore more expensive than in other industries.

5   By contrast, the process is slightly faster in ferrous metallurgy. This is partly due to the fact that production processes in this branch are quite similar across enterprises. Moreover, companies even use equipment from the same manufacturers.

6   This is demonstrated by the rates of implementation of similar solutions at **plants** operated by the leading global manufacturers, including Tata Steel and Severstal. Thus, one of the enterprises uses industrial **IoT** and AI technologies to create a recommendation system for the **electric arc furnace** (EAF) operator. Without upgrading the current production equipment, this system reduces time spent in the energized state during each smelting operation with no loss of quality (**incl.** to FeO content in slag, phosphorus and nitrogen content in the semi-finished product).

7   These systems, based on IoT and AI technologies, are being implemented not only in smelting processes but also in other production cycles.

8   AI is widely used where the traditional automation tools have been unable to challenge the expert metallurgist — in operational solutions with **fuzzy** logic. As a rule, these are routine decisions which employees have to make daily or sometimes even dozens of times a day.

9   At its current level of development, AI is not yet capable of creating fundamentally new methods or solutions. Its function is rather **optimization**: reducing the volume of consumables or energy, searching for **optimal** equipment operating modes, and quality control.

10   Since this means optimization of the established production processes, transition to a fundamentally new efficiency level through AI implementation should not be **anticipated**: as a rule, the impact varies from 3% to 10%. A strong efficiency gain shows that the process was initially not set up in the best way and the result is more likely **integral**. But even an optimization of 3% of expenses is financially significant for large and medium enterprises.

11   The **empirical** experience has shown that the use of AI optimization models for the **flotation** process increases the extraction factor by an average of 3.5%, while in **electrolysis** the efficiency gain is 4.7% etc.

12   "**Machine vision**" is one of the fastest-growing and most popular areas of application of artificial intelligence in nonferrous metallurgy. It is focused on the development of technologies to create images of real-world objects, process and analyse them, and then use the data obtained to solve applied problems. For instance, for monitoring of a specific production area: online tracking of cathode and anode closing in the electrolysis shop, recognizing the material on the conveyor and classifying it according to quality.

13　Solutions based on this technology have already been implemented at Rusal Group enterprises in **electrolytic** equipment that is already operating. The use of "machine vision" technologies in electrolysis shops enables not only production process optimization and significant cost reductions (incl. on electricity consumption), but also enhances production safety and environmental friendliness.

14　"**Digital twin**" is the creation of a mathematical model of a production process. Production operators can use data from "digital twins" to receive real-time recommendations on production process control. The solution is used to monitor the consumption of materials and energy resources and detect **reject** items, as well as to monitor process **parameters**. Thus, at Kola Mining and Metallurgical Company, the "digital twin" optimizes **matte** delivery from the smelting furnaces to the converters.

15　At present, AI is often still recommendatory in most applications, though some systems with fully automatic control based on AI technologies are available. However, **personnel** are still essential for the **supervisory** function, as well as when manual control is required.

16　In other words, even 20 years from now there will be no completely unmanned enterprises, though this concept has been under discussion for many years. It will still be impossible to do without **maintenance** personnel. By contrast, many operations, including hazardous ones, will be performed by robots instead of humans at a certain stage of AI development in this field, when a **sufficient** mass of experimental data has been accumulated. And this is reasonable, because labour and health safety standards in metallurgical production are very high. Accordingly, if humans in some areas with harmful working conditions can be **substituted** with robots performing the necessary operations more quickly, why not do this?

17　Thus, there will still be a demand for metallurgical engineers, but ones who are both knowledgeable about process theory and have a clear understanding of the business system, as well as programmer engineers to create the artificial intelligence which will control all plant processes.

◆ Words and Expressions

auxiliary /ɔːɡˈzɪliəri/ *adj.* (of a piece of equipment) used if there is a problem with the main piece of equipment 备用的

accelerate /əkˈseləreɪt/ *v.* to happen or to make something happen faster or earlier than expected (使)加速；加快

per annum /pər ˈænəm/ *adv.* for each year 每年

subdivision /ˈsʌbdɪvɪʒn/ *n.* one of the smaller parts into which a part of something has been

Unit 6 Development of Nonferrous Metallurgy

divided 进一步分成的部分；分支；分部

monetization /ˌmʌnɪtaɪˈzeɪʃn/ n. the act or process of earning money from something, especially a business or an asset 货币化

implement /ˈɪmplɪmnt/ v. to make something that has been officially decided start to happen or be used 使生效；贯彻；执行；实施

slightly /ˈslaɪtli/ adv. a little 略微；稍微

prospect /ˈprɒspekt/ n. the chances of being successful 成功的机会；前景；前途

content /ˈkɒntent/ n. the amount of a substance that is contained in something else 含量；容量

admixture /ədˈmɪkstʃə/ n. a mixture 混合；掺和

feedstock /ˈfiːdstɒk/ n. the raw material that is used to produce something in an industrial process 原料；给料

intermediate /ˌɪntəˈmiːdiət/ v. having more than a basic knowledge of something but not yet advanced; suitable for somebody who is at this level 中级的；中等的；适合中等程度者的

favourable /ˈfeɪvərəbl/ adj. good for something and making it likely to be successful or have an advantage, advantageous 有利的；有助于……的

spectrum /ˈspektrəm/ n. [usually singular] a complete or wide range of related qualities, ideas, etc. 范围；各层次；系列

upscale /ˈʌpskeɪl/ v. to make something better, bigger or more powerful 改进；扩大；升级

plant /plɑːnt/ n. a factory or place where power is produced or an industrial process takes place 发电厂；工厂

IoT (abbreviation) 物联网 (Internet of Things)

incl. (abbreviation) including; included 包括；连同……在内

fuzzy /ˈfʌzi/ adj. not clear in shape or sound, blurred（形状或声音）模糊不清的

optimization /ˌɒptɪmaɪˈzeɪʃən/ n. the act of making something as good as possible 优化

optimal /ˈɒptɪməl/ adj. the best possible; producing the best possible results 最佳的；最适宜的

anticipate /ænˈtɪsɪpeɪt/ v. to expect something 预料；预期

integral /ˈɪntɪɡrəl/ adj. being an essential part of something 必需的；不可或缺的

empirical /ɪmˈpɪrɪkl/ adj. (formal) based on experiments or experience rather than ideas or theories 以实验（或经验）为依据的；经验主义的

flotation /fləʊˈteɪʃən/ n. the act of floating on or in water 浮；漂浮

reject /rɪˈdʒekt/ n. something that cannot be used or sold because there is something wrong with it 废品；次品

parameter / pəˈræmɪtə/ *n.* (formal) something that decides or limits the way in which something can be done 决定因素；规范；范围

personnel / pɜːsəˈnel/ *n.* the people who work for an organization or one of the armed forces (组织或军队中的)全体人员；职员

supervisory / ˌsuːpəˈvaɪzəri/ *adj.* in charge of people or activities 监督的；指导的

maintenance / ˈmeɪntənəns/ *n.* the act of keeping something in good condition by checking or repairing it regularly 维护；保养

sufficient / səˈfɪʃənt/ *adj.* enough for a particular purpose; as much as you need 足够的；充足的

substitute / ˈsʌbstɪtjuːt/ *v.* to take the place of somebody/something else; to use somebody/something instead of somebody/something else (以……)代替；取代

take into account to consider particular facts, circumstances, etc. when making a decision about something 考虑到；顾及

◆ Terminology

zinc / zɪŋk/ *n.* 锌

lead / liːd/ *n.* 铅

electrolysis / ˌɪlekˈtrɒləsɪs/ *n.* 电解

electrolytic / ɪˌlektrəʊˈlɪtɪk/ *adj.* 电解质的

matte / mæt/ *n.* 锍

electric arc furnace (EAF) 电弧炉

machine vision (MV) 机器视觉

digital twin 数字孪生

Unit 6 Development of Nonferrous Metallurgy

Section D Language Focus

Task One | Text organization
Work in groups and discuss the organization of the text and fill in the blanks.

Parts	Paragraphs	Main Ideas
Part 1	Para. 1-2	Introduction to the 1. _____ of artificial intelligence at present
Part 2	Para. 3-7	Digitization 2. _____ in nonferrous metallurgy
Part 3	Para. 8-11	Artificial intelligence 3. _____ helpful for enterprises profit
Part 4	Para. 12-14	Promising developments: 4. _____ vision and 5. _____ twin
Part 5	Para. 15-17	6. _____ of artificial intelligence

Task Two | Answer the following questions based on the information contained in the text.

1. How much percent will the industrial AI market grow, according to forecasts by Technavio? _____

   A. More than 54% per month.

   B. More than 54% per year.

   C. 54% a month on average.

   D. 54% a year on average.

2. What is the synonym of the word "application" in Paragraph 4? _____

   A. Efficiency.

   B. Implementation.

   C. Product.

   D. Process.

3. What is the function of AI in nonferrous metallurgy? _____

   A. Creating new methods.

   B. Creating new solutions.

   C. Optimizing the established production processes.

   D. Searching for equipment.

4. According to the passage, which is **NOT** a characteristic of machine vision? _____

   A. Machine vision is one of the widely used areas of application of artificial intelligence in nonferrous metallurgy.

   B. The use of "machine vision" technologies in electrolysis shops is bad for environmental friendliness.

C. Machine vision is useful to create pictures of real-world objects, process and analyse them, and then use the data to solve problems.

D. Solutions based on machine vision have already been used at Rusal Group enterprises in electrolytic equipment.

5. What does the writer believe? _____

A. Even though AI technologies are recommended, personnel are not that important.

B. Twenty years from now there will be unmanned enterprises.

C. Robots can take place of humans in some areas with dangerous working conditions.

D. Metallurgical engineers should be both knowledgeable about process theory and have a good understanding of the business system.

Task Three | Fill in the blanks with the words or phrases given in the box. You may not use any of the words or phrases more than once.

Machine vision systems are a set of integrated 1. _____ that are designed to use information 2. _____ from digital images to automatically guide manufacturing and production operations such as go/no testing and quality 3. _____ processes. These systems can also play a 4. _____ in automated assembly verification and inspection operations through their ability to guide material handling equipment to position products or materials as 5. _____ in a given process. They have wide applications across different industries and can be used to automate any mundane, 6. _____ tasks that would become tiring to a human 7. _____ or operator. The use of machine vision systems 8. _____ for 100% inspection of products or parts in a process, 9. _____ in improved yields, reductions in 10. _____ rates, increased quality, lower costs, and greater consistency of process results.

A. control	I. resulting
B. defect	J. needed
C. repetitive	K. worker
D. components	L. extracted
E. worked	M. accounts
F. role	N. delivered
G. false	O. allows
H. inspector	

Unit 6 Development of Nonferrous Metallurgy

Task Four | The following passage has four paragraphs. Choose the correct topic sentence for each paragraph from the list of sentences below.

A. Because of the toxicity of lead, traditional uses like paint and water pipes have now declined.
B. Throughout the world the automotive industry is thriving, and lightweight, nonferrous metals are a large part of that.
C. Globally, the last few years have been tempestuous for the nonferrous metal industry with an economic slowdown and high raw material costs.
D. We've come a long way since the Copper and Bronze Ages, and the many uses of nonferrous metals permeate the world around us.
E. Aluminium is very strong, rigid, light, and also has many environmental advantages.

1. _____ With their light weight, non-magnetic characteristics, and a higher resistance to corrosion and rust than ferrous metals, they are found in anything from jewellery and electronic applications to construction and the automotive industry.

2. _____ Aluminium is increasingly useful bearing in mind the rigorous laws relating to emissions and fuel efficiency. As a light metal, it is an excellent replacement for the traditional steel, ensuring manufacturers reduce the weights of their vehicles with ease.

3. _____ However, it is still used for lead-acid car batteries to great effect.

4. _____ Its strength means it can absorb twice the crash energy of an equivalent structure made out of mild steel, while its lightweight capabilities mean the vehicle uses less fuel. The recycling possibilities are endless as 90% of a vehicle's aluminium can be recycled with very little degradation. It is used to make heat exchangers, wheels, heads, blocks, brake and suspension components, and some chassis, many of which can be diecast.

Task Five | Translate the following sentences into English, using the words or phrases in the brackets.

1. 中国的有色金属储量丰富，品种繁多，有"有色金属王国"之称。(nonferrous metals)

2. 在熔炉里的 2000 吨熔融玻璃中，若干过程——熔化、精炼、均质化——在同时进行。（homogenizing, furnace）

3. 在形成的过程中，电子可以和水分子相互作用，并在电解过程中将它们分开。（electron, electrolysis）

4. 随着欧洲用市政焚化炉取代垃圾填埋场后，大量的炉渣产生，里面含有少量的金属。（municipal incinerator, slag）

5. 人们已经发现，钠电池能产生更大的能量，其电动马达强度是锌电池的两倍。（zinc）

Unit 6 Development of Nonferrous Metallurgy

Part II: Text B

Section A Warm-up Questions

Discuss with your partners about the following questions.
1. Do you know microbes? What are examples of microbes?
2. How can microbes be helpful?

Section B Listening Practice

Listen to the audio clips and fill in the blanks with missing information.

In Wales, in Derbyshire, and in the north of England, hills made of limestone are very common. In the cracks of this limestone we often find a beautiful silvery mineral called galena. It is made of lead 1. _____.

Minerals or stones, like galena, which contain useful metals, are known as ores. To get the lead from it, this ore is first 2. _____ and roasted in a furnace at a gentle heat. Then the furnace is made hotter and hotter, until the lead melts and runs off.

Lead is between blue and grey in colour, so we call it bluish-grey. It is so soft that it can be easily 3. _____. If we scrape a piece of lead with a knife, the new surface is bright and shining. But in a few days it gets dim or rusts by being exposed to the air.

Men who work with lead are called plumbers. If you watch a plumber at work, you will see how easily he 4. _____, and how cleverly he hammers the sheet-lead with a mallet, and makes it fit into any corner. Because lead bends so easily, we call it a flexible metal.

Lead can be hammered or rolled out into thin sheets. When the sheets are very thin, they are called lead-foil. A great deal of lead-foil is used in China, to 5. _____ in which tea is sent to Europe.

Section C Active Reading

Biomining: How Microbes Help to Mine Copper

1 Say bacteria. People think infection. Or yogurts. But in Chile, bacteria are being used to get at something this country heavily depends on: copper.

2 Chile is the world's biggest copper exporter, and has the planet's largest known reserves of the red metal. The Atacama Desert, a **desolate** rocky plateau west of the Andes mountains, is **dotted** with copper mines. Every now and then, a mining train passes through the arid landscape. Here and there the ruins of tiny towns still stand where miners used to live decades ago, their mud huts gradually falling to pieces under the **baking** sun. Export of the metal is essential for Chile's economy—it **amounts to** about 70% of all Chilean exports—and the more copper the country digs out, the more money pours in.

3 The demand for the metal is continually increasing too; copper is used in most of our lives, from electrical wires and telephone lines to roofing materials, from nutritional supplements to jewellery. It is thought that the element was created billions of years ago in the stars and then became part of the materials that formed the Earth. Volcanoes and other activity caused by the planet's shifting **tectonic** plates then brought it closer to the surface.

4 To get it out, you normally have to dig. Then, to separate copper from ore-the rocks containing it—you have to crush and grind, and then apply sweltering heat and toxic chemicals. These conventional mining methods are very energy-intensive and thus expensive, and therefore only used on sites where it is estimated that the **concentration** of copper—called grade—is actually worth spending that much money on its extraction.

Bacteria on board

5 But as people have been hunting for the Earth's minerals for centuries, high-grade copper sites have become **scarce**. In the past there were deposits that contained as much as 30% of the red metal. Now many have grades of 1% to 1.8%. But even when a mine is estimated to contain high-grade ores, the **bulk** of the material coming out of the pit has a grade below 1%, and is usually discarded as waste. Unless mini-miners come to help-microbes.

6 Biosigma is a biotechnology **venture** set up by Codelco, a **Chilean** state-owned corporation and the largest copper mining company in the world, together with Japanese firm Nippon Metals and Mining. Based on the **outskirts** of Santiago, Biosigma is one of only a **handful** of biomining companies around the world.

7   In the main lab, chemists and biologists in white coats and **goggles** are busy transferring colourful liquids from **flasks** to tubes and performing peculiar tests in front of an open oven with **raging** fire.

8   All the flasks, tubes, containers and huge tanks are full of microbes: **Acidithiobacillus ferrooxidans** and **Thiobacillus** ferrooxidans bacteria, harnessed by the firm to break down minerals in order to improve copper recovery rates and reduce operating costs.

9   "We know that conventional mining methods are not used for low-grade materials that simply get dumped—so the only way to get copper from them is by using new knowledge and capacities—in this case, biotech," says Ricardo Badilla, chief executive of Biosigma.

10   Using bacteria can result in extracting as much as 90% of the total metal at a pit mine, instead of merely 60%, he adds.

11   But how can microbes help get copper from a rock?

12   As Pilar Parada, research and development director of Biosigma, walks past a weird-looking **installation** with wildly shaking flasks, she says the key to success is using **microorganisms** that are naturally present at mining sites. "These bacteria need very little to do their work, they use air and mainly oxygen and CO_2, and use the mineral itself as a source of energy," she says. She explains that if a mining site is left alone, microorganisms would eventually liberate copper from rocks, but it could take hundreds of years.

13   To speed up the process in a biomining lab, scientists use bioleaching. Ores are placed into acid, and then researchers introduce bacteria that change the solution so that it **dismantles** the rock and frees copper, in liquid form. And after a special electrochemical process, it is then turned into solid metal that can be used in the industrial applications we so much depend on.

Future of mining

14   Some call biomining the "mining of the future". Indeed, it is much cheaper and greener than traditional mining—there are a lot fewer CO_2 emissions and carbon and water footprints are lower than using conventional technology. Furthermore, the toxic chemicals used in traditional mining can be extremely harmful to the environment; there have been accidents before. In the case of biomining, the bacteria are naturally occurring at mining sites anyway, and are not **pathogenic**.

15   Biomining is already in use in several countries, including South Africa, Brazil and Australia. Overall, some 20% of the world's copper production comes from bioleaching. The practice is not limited to copper. Microorganisms are also used to extract gold and **uranium**. And there are other applications of biomining: scientists are working on using microbes to clean up the corrosive acid pollution left over in mining waste. But Gabriel Rodriguez, the director for energy, science, technology and innovation at Chile's Ministry of Foreign Affairs, says that the technology needs help

to develop. "There are still not enough microorganisms for doing that job, so more research needs to be done," he says. "And this is exactly the bet that Chile has been making in the last years. But our bet is also to export the technology, so that the world can move from just exploiting natural resources to adding a value with the help of biotech."

16 If it works, one day it might be possible to get mine for copper without digging huge pit mines. Instead, miners would simply drill two holes to introduce a solution full of microbes, and then collect it once it contains copper. This alternative to traditional mining could even help save lives—estimates suggest that some 12000 people die from mining accidents around the world each year.

17 The 33 miners who spent 69 days trapped at 700m below the surface at a copper and gold mine in Chile's Copiapo region in 2010 had a lucky escape; not everyone is that lucky. If microbes take over, the bacteria won't die if they get trapped.

◇ Words and Expressions

desolate / ˈdesələt/ *adj.* empty and without people, making you feel sad or frightened 无人居住的；荒无人烟的；荒凉的

dot / dɒt/ *v.* to spread things or people over an area; to be spread over an area 星罗棋布于；遍布

baking / ˈbeɪkɪŋ/ *adj.* extremely hot 灼热的；炽热的

scare / skeəs/ *adj.* if something is scarce, there is not enough of it and it is only available in small quantities 缺乏的；不足的；稀少的

bulk / bʌlk/ *n.* the main part of something; most of something 主体；大部分

venture / ˈventʃə/ *n.* a business project or activity, especially one that involves taking risks （尤指有风险的）企业；商业；经营项目

Chilean / ˈtʃɪliən/ *adj.* from Chile 智利的；智利人的

outskirts / ˈaʊtˌskɜːts/ *n.* the parts of a town or city that are furthest from the centre （市镇的）边缘地带；市郊

handful / ˈhændfʊl/ *n.* a small number of people or things 少数人（或物）

goggles / ˈgɒglz/ *n.* a pair of glasses that fit closely to the face to protect the eyes from wind, dust, water, etc. 护目镜；风镜；游泳镜

flask / flɑːsk/ *n.* a bottle with a narrow top, used in scientific work for mixing or storing chemicals 烧瓶

raging / ˈreɪdʒɪŋ/ *adj.* very powerful 极其强大的；猛烈的

Unit 6　Development of Nonferrous Metallurgy

installation /ˌɪnstəˈleɪʃn/ *n.* a piece of equipment or machinery that has been fixed in position so that it can be used 安装的设备(或机器)；装置

dismantle /dɪsˈmæntl/ *v.* to take apart a machine or structure so that it is in separate pieces 拆开；拆卸(机器或结构)

amount to to add up to something; to make something as a total 总计；共计

◆ Terminology

microbe /ˈmaɪkrəʊb/ *n.* 微生物
tectonic /tekˈtɒnɪk/ *adj.* 地壳构造的
concentration /ˌkɒnsnˈtreɪʃn/ *n.* 浓度；含量
acidithiobacillus /əsidiθaiəubəˈsiləs/ *n.* 酸硫杆菌
ferrooxidans /ferəʊˈɔksidəns/ *n.* 氧化铁
thiobacillus /θaiəubəˈsiləs/ *n.* 硫杆菌；产硫酸杆菌
microorganism /ˌmaɪkrəʊˈɔːɡənɪzəm/ *n.* 微生物
pathogenic /ˌpæθəˈdʒenɪk/ *adj.* 致病的；病原的；发病的
uranium /jʊˈreɪniəm/ *n.* 铀(放射性化学元素)

☞ Section D　Language Focus

Task One ｜ Are the following statements True or False according to the passage? Write T/F accordingly.

1. Chile is a country that greatly depends on copper.　　　　　　　　　(　　)
2. Since the conventional mining methods are used on sites, they are very expensive. (　　)
3. The key to getting copper from a rock for high-grade materials is using microorganisms.
　　　　　　　　　　　　　　　　　　　　　　　　　　　　　　　　(　　)
4. The biomining is much cheaper and greener than conventional mining. (　　)
5. Bioleaching is only limited to copper production.　　　　　　　　　　(　　)

Task Two | Translate the following sentences into Chinese.

1. Chile is the world's biggest copper exporter, and has the planet's largest known reserves of the red metal. The Atacama Desert, a desolate rocky plateau west of the Andes mountains, is dotted with copper mines.

2. But as people have been hunting for the Earth's minerals for centuries, high-grade copper sites have become scarce. In the past there were deposits that contained as much as 30% of the red metal.

3. All the flasks, tubes, containers and huge tanks are full of microbes: Acidithiobacillus ferrooxidans and Thiobacillus ferrooxidans bacteria, harnessed by the firm to break down minerals in order to improve copper recovery rates and reduce operating costs.

4. To speed up the process in a biomining lab, scientists use bioleaching. Ores are placed into acid, and then researchers introduce bacteria that change the solution so that it dismantles the rock and frees copper, in liquid form.

Unit 6 Development of Nonferrous Metallurgy

5. If it works, one day it might be possible to get mine for copper without digging huge pit mines. Instead, miners would simply drill two holes to introduce a solution full of microbes, and then collect it once it contains copper.

Task Three | Choose the best answers for each blank of the following passage.

How do you break down kilometers of 1._____ to get the metals within? Most of the time, you would smash it and dunk it in powerful 2._____ until everything dissolves, then sieve out the metal. 3._____ you could get someone who could make acid where it was needed, without compromising the environment? Like a worker... who gets paid in water.

Bioleaching uses 4._____ that make their own organic acids to break down minerals. As the bacteria feed, they 5._____ rare-earth metals, which run off in the water fed through the rock. Because the valuable metals are in the run-off, the bioleaching system is contained, meaning 6._____ pollution than traditional methods.

7._____ bioleaching for rare-earth metals is new, the method is already being used to mine copper in Western Australia and around the world. These bioleaching piles are giant 8._____ reactors that break down rocks for energy. They can be thousands of kilometers across. Water is drip filtered through the rock, 9._____ the bacteria to grow and letting the metals collect in the run-off. One local species used for bioleaching was actually found in an old open-cut coal mine in Collie, Western Australia.

While the acidic, toxic environment doesn't seem like a 10._____ home, it was an all-you-can-eat-buffet for Alicyclobacillus acidocaldarius. These bacteria have 11._____ to use those elements as part of their metabolism—much like the mineral-munching microbes found around volcanic vents under the sea.

"In this space, we work with iron and sulfur-oxidizing bacteria," says Elizabeth, a researcher in molecular microbial ecology. "When iron and sulfur are 12._____, they release electrons which the bacteria use for ATP [energy] production."

In the wild, these bacteria are 13._____, releasing their acid into nearby water systems. But when 14._____, they may actually prove more environmentally friendly than the artificial acids 15._____ jobs they're taking.

1. A. element   B. stone   C. rock   D. crust
2. A. acid   B. water   C. liquid   D. bacteria
3. A. Even if   B. What if   C. As if   D. So what
4. A. acid   B. liquid   C. water   D. bacteria
5. A. give   B. expose   C. send   D. move
6. A. more   B. much   C. less   D. little
7. A. While   B. Because   C. If   D. When
8. A. alive   B. live   C. living   D. lively
9. A. making   B. allowing   C. letting   D. moving

Unit 6 Development of Nonferrous Metallurgy

10. A. cozy	B. cold	C. hot	D. cool
11. A. involved	B. reached	C. used	D. evolved
12. A. joined	B. mixed	C. dunk	D. oxidized
13. A. helpful	B. harmful	C. useful	D. meaningful
14. A. ruled	B. ruling	C. controlled	D. controlling
15. A. whose	B. who	C. which	D. that

Task Four | The following passage has five paragraphs. Choose the correct topic sentence for each paragraph from the list of sentences below.

> A. Biomining is so versatile that it can be used on other planetary bodies.
> B. The source of energy required depends on the specific microbe necessary for the job.
> C. It sounds futuristic, but it's currently used to produce about 5% of the world's gold and 20% of the world's copper.
> D. All living organisms need metals to carry out basic enzyme reactions.
> E. Biomining takes place within large, closed, stirred-tank reactors (bioreactors).

Biomining is the kind of technique promised by science fiction: a vast tank filled with microorganisms that leach metal from ore, old mobile phones and hard drives.

1. _____ It's also used to a lesser extent to extract nickel, zinc, cobalt and rare earth elements. But perhaps it's most exciting potential is extracting rare earth elements, which are crucial in everything from mobile phones to renewable energy technology.

2. _____ Bioleaching studies on the international space station have shown microorganisms from extreme environments on Earth can leach a large variety of important minerals and metals from rocks when exposed to the cold, heat, radiation and vacuum of space.

3. _____ These devices generally contain water, microorganisms (bacteria, archaea, or fungi), ore material, and a source of energy for the microbes.

4. _____ For example, gold and copper are biologically "leached" from sulfidic ores using microorganisms that can derive energy from inorganic sources, via the oxidation of sulfur and iron.

Part III: Academic Skills

Academic Writing Skills (VI) *

1. Academic Writing in Different Disciplines

In terms of academic writing, general advice is offered to support your development in terms of textual investigation and academic writing but each discipline has its own conventions and it is important to take careful note of these in your writing. The lectures in your discipline and the texts you read in each discipline are the most important source of information for discipline-specific writing. The focus so far has been on mastering academic writing but there are disciplinary differences which are important for you to understand so that you can produce good work.

This section is to help you to explore writing practices in your subject area or discipline. In order to write well, you need to approach writing tasks with an understanding that writing is discipline specific. Writing in the disciplines varies widely in terms of content, research methods and citation styles. The citation style in Humanities is very different from Science and Engineering. The use of the passive voice is preferred in almost all cases but there are times when the use of "I" is acceptable.

To develop your academic writing, you will need to express your own thoughts or views on the material; you cannot rely on the ideas and thoughts of other people. However, in the early years of undergraduate study your contribution will relate to the choice you make about what literature you want to present and to how you analyse it. In the Humanities and Social Sciences, what is important is that your views should be informed, clearly expressed and based on careful consideration of the views of seminal writers and thinkers on the topic. In scientific disciplines, you must show that you have a complete knowledge and understanding of the relevant scientific principles, but in either discipline it is in the analysis and interpretation that you can make your contribution.

The aim of the activity is to help you become familiar with the style of writing in your discipline so that you can, from the beginning, approach your writing in the way required in your discipline.

---

* Taken from *Developing Your Academic Writing Skills: A Handbook* by Marian Fitzmaurice & Ciara O'Farrell.

Activity: Read and compare two peer reviewed journal articles from your discipline on a topic you are researching and answer the following questions.

	Article 1	Article 2
What is the length of the introductions?		
Do the articles use any headings or sub-headings?		
What is the style of referencing used?		
What is the typical paragraph length?		
How many long quotations are used?		
Does the text use more paraphrasing than quotation?		
Create a list of common verbs used to refer to outside sources and to introduce quotations (e. g. according to, argues that, describes, concludes).		
Does the author use the first person pronoun?		
Does the author make use of diagrams or any other type of visuals?		
What types of evidence is used in making argument?		

2. Summary Guidelines on Academic Writing

• Write every day as writing is a generative process and putting pen to paper helps you to think more clearly.

• Start to write early and don't postpone and procrastinate as there is never a perfect time.

• Keep your topic in mind and do some freewriting exercises.

• Break a large piece of writing into manageable pieces; headings can be useful here.

• Write first and edit/revise later as they are different activities.

• Remember that references are a tool to help you to make a point or develop an argument but you must acknowledge your sources.

• Ensure all citations and referencing are correct so that the reader understands how sources have been used.

• Draft and redraft as you will not get it all right the first time.

● Remember that there are stages to writing: planning, freewriting, writing drafts, revising, editing.

● Readers need a route map to guide them through the work, so write a good introduction to make it clear what they are about to read.

Unit 6 Development of Nonferrous Metallurgy

Part Ⅳ: Extended Reading and Translation

Translate the following paragraph into English.

考古学家宣布，四川广汉三星堆遗址新发现的6座三星堆文化"祭祀坑"中已出土500余件文物。此次发现的"祭祀坑"距今约3200多年。出土的重要文物包括金面具、铜面具、铜器、象牙、纺织物以及玉器等其他工艺品。在三号坑，考古学家发现了大量青铜器，其中有两个正方形青铜尊，是中国古代典型的青铜祭器。其中一些青铜器上面还装饰有龙、牛的图案。这些文物都是1986年发掘时没有发现的。这次的发现进一步证明，这些坑洞是为祭祀所用，因为出土的很多文物在埋之前已经被打碎并烧毁。需要更深入的研究来排除这些坑洞有其他用途的猜想。

Glossary

A

weaponry /ˈwepənri/ *n.* 武器，兵器	U1TA
accelerate /əkˈseləreɪt/ *v.* （使）加速；加快	U6TA
accumulate /əˈkjuːmjəleɪt/ *v.* 积累；积聚	U3TB
acidify /əˈsɪdɪfaɪ/ *v.* （使）变成酸，酸化	U5TB
additive /ˈædətɪv/ *n.* （尤指食品的）添加剂，添加物	U3TB
additive manufacturing 增材制造；添加剂制造	U5TA
adjoin /əˈdʒɒɪn/ *v.* 紧挨；邻接；毗连	U5TB
adjoining /əˈdʒɔɪnɪŋ/ *adj.* 邻接的；毗连的	U1TA
admixture /ədˈmɪkstʃə/ *n.* 混合；掺和	U6TA
adsorption /ædˈsɔːpʃən/ *n.* 吸附（作用）	U4TB
advent /ˈædvənt/ *n.* （重要事件、人物、发明等的）出现，到来	U5TA
aerospace /ˈeərəspeɪs/ *n.* 航空航天工业	U5TA
age hardening 时效硬化	U4TA
agglomerate /əˈɡlɒməreɪt/ *v.* （使）成团，聚结	U1TA
alternative /ɔːlˈtɜːnətɪv/ *adj.* 可供替代的	U1TB
ambient temperature 常温	U5TB
amorphous /əˈmɔːfəs/ *adj.* 不规则的	U3TB

Glossary

amount to 总计；共计   U6TB

anneal /ə'niːl/ v. 给(金属或玻璃)退火   U4TA

anneal /ə'niːl/ v. 使退火   U3TA

anode /'ænəʊd/ n. 阳极；(电解池的)正极；(原电池的)负极   U4TB

anticipate /æn'tɪsɪpeɪt/ v. 预料；预期   U6TA

applicable /ə'plɪkəbl/ adj. 适用的   U3TB

aqueous /'eɪkwɪəs/ adj. 水的；含水的；水状的   U4TB

arbitrariness /'ɑːbɪ'trərinis/ n. 任意；武断；随心所欲   U4TB

array /ə'reɪ/ n. 大堆；大群；大量   U5TA

assemble /ə'sembl/ v. 装配；组装   U2TA

assume /ə'sjuːm/ v. 假定；假设；认为   U4TA

atomization /ˌætəʊmaɪ'zeɪʃən/ n. 雾化，[分化] 原子化   U5TA

attain /ə'teɪn/ v. (通常经过努力)获得；得到   U4TA

austenitic /ˌɔːstə'nɪtɪk/ adj. 奥氏体的   U3TA

auxiliary /ɔːg'zɪlɪərɪ/ adj. 备用的   U6TA

azurite /'æʒʊraɪt/ n. 蓝铜矿   U5TB

B

baking /'beɪkɪŋ/ adj. 灼热的；炽热的   U6TB

ballast /'bæləst/ n. (用作公路或铁路路基的)道砟   U3TB

base metal 贱金属   U5TA

bauxite /'bɔːksaɪt/ n. 铝土矿   U1TB

bearing /'beərɪŋ/ n. (机器的)支座；(尤指)轴承   U5TA

bellows /'beləʊz/ n. 风箱；吹风器   U2TB

beryllium /bə'rɪlɪəm/ n. 铍   U4TA

Bessemer process（旧）酸性转炉炼钢法　　　　　　　　　　　　　　　　　　　　U1TB

binding agent［胶粘］粘合剂；黏合剂　　　　　　　　　　　　　　　　　　　　U5TA

bituminous /bɪˈtjuːmɪnəs/ *adj.* 含沥青的；沥青的　　　　　　　　　　　　　　　U1TB

blast furnace /ˈblɑːstˈfɜːnɪs/ *n.* 鼓风炉；高炉　　　　　　　　　　　　　　　　U1TA

blister steel /ˈblɪstə(r) stiːl/［冶］泡钢，泡面钢　　　　　　　　　　　　　　　　U1TA

bloomery /ˈbluːmərɪ/ *n.* 锻铁炉；熟铁块吹炼法　　　　　　　　　　　　　　　U2TB

bog ore /bɒg ɔː(r)/ *n.* 沼矿　　　　　　　　　　　　　　　　　　　　　　　　U2TB

bosh /bɒʃ/ *n.*（高炉风嘴以上的）炉腹　　　　　　　　　　　　　　　　　　　　U2TB

brake /breɪk/ *n.* 制动器；车闸　　　　　　　　　　　　　　　　　　　　　　　U2TA

brittleness /ˈbrɪtlnəs/ *n.* 脆性　　　　　　　　　　　　　　　　　　　　　　　U1TA

brochantite /ˌbrɔtʃænˈtaɪt/ *n.*［矿物］水胆矾　　　　　　　　　　　　　　　　U5TB

bulk /bʌlk/ *n.* 主体；大部分　　　　　　　　　　　　　　　　　　　　　　　　U6TB

burnt lime 煅石灰；氧化钙　　　　　　　　　　　　　　　　　　　　　　　　　U3TB

burr /bɜː(r)/ *v.*（机器部件快速运转时有规律的）呼呼声　　　　　　　　　　　　U2TA

C

candidate /ˈkændɪdət/ *n.* 有望做…的人；有望成为…的事　　　　　　　　　　　U3TB

cast iron /ˌkɑːst ˈaɪən/ *n.* 铸铁　　　　　　　　　　　　　　　　　　　　　　U1TA

cathode /ˈkæθəʊd/ *n.* 阴极；（电解池的）负极；（原电池的）正极　　　　　　　U4TB

cavity /ˈkævəti/ *n.* 洞；孔　　　　　　　　　　　　　　　　　　　　　　　　U2TA

cementation /siːmenˈteɪʃn/ *n.*（金属的）渗镀　　　　　　　　　　　　　　　　U1TA

cemented carbide［材］硬质合金；烧结硬质合金　　　　　　　　　　　　　　　U5TA

ceramic /səˈræmɪk/ *n.* 陶瓷制品；陶瓷器　　　　　　　　　　　　　　　　　　U3TB

chalcocite /ˈkælkəsaɪt/ *n.*［矿物］辉铜矿　　　　　　　　　　　　　　　　　　U5TB

chalcopyrite /ˌkælkəˈpaɪraɪt/ *n.*［矿物］黄铜矿　　　　　　　　　　　　　　　U5TB

Glossary

chamber /ˈtʃeɪmbə(r)/ n. （作特殊用途的）房间，室 　　U1TB

charcoal /ˈtʃɑːkəʊl/ n. 炭，木炭(可作燃料或供作画) 　　U1TA

Chilean /ˈtʃɪliən/ adj. 智利人；智利的；智利人的 　　U6TB

chromium /ˈkrəʊmiəm/ n. 铬 　　U1TB

chrysocolla /ˌkrɪsəˌkɒlə/ n. [矿物] 硅孔雀石 　　U5TB

clay /kleɪ/ n. 黏土；陶土 　　U1TB

coefficient /ˌkəʊɪˈfɪʃnt/ n. 系数 　　U3TB

collier /ˈkɒliə(r)/ n. 煤矿工人 　　U2TB

composition /ˌkɒmpəˈzɪʃn/ n. 成分；构成 　　U3TB

compound /ˈkɒmpaʊnd/ n. 复合物；混合物 　　U1TA

compressive /kəmˈpresɪv/ adj. 压缩的；有压缩力的 　　U2TA

Computer Numerical Control 电脑数控 　　U2TA

concentrate /ˈkɒnsntreɪt/ n. 浓缩物 　　U5TB

concentration /ˌkɒnsnˈtreɪʃn/ n. 浓度；含量 　　U6TB

concern /kənˈsɜːn/ n. （尤指许多人共同的）担心，忧虑 　　U3TB

conductivity /ˌkɒndʌkˈtɪvəti/ n. 导电性 　　U3TB

connector /kəˈnektə(r)/ n. 连接物；连接器；连线 　　U4TA

contaminant /kənˈtæmɪnənt/ n. 致污物；污染物 　　U5TB

content /ˈkɒntent/ n. 含量；容量 　　U6TA

conventional /kənˈvenʃənl/ adj. 传统的；习惯的 　　U5TB

converter /kənˈvɜːtə(r)/ n. 转炉 　　U1TB

corrosion /kəˈrəʊʒən/ n. 腐蚀 　　U3TB

corrosive /kəˈrəʊsɪv/ adj. 腐蚀性的；侵蚀性的 　　U5TA

counterpart /ˈkaʊntəpɑːt/ n. 职位(或作用)相当的人；对应的事物 　　U5TB

covellite /kəʊˈvelaɪt/ n. [矿物]铜蓝；靛铜矿　　　　　　　　　　　　U5TB

craft /krɑːft/ v. （尤指用手工）精心制作　　　　　　　　　　　　　U2TA

crease /kriːs/ n. 褶痕；皱痕　　　　　　　　　　　　　　　　　　U2TA

crucible /ˈkruːsɪbl/ n. 熔炉　　　　　　　　　　　　　　　　　　U1TA

crystal /ˈkrɪstl/ adj. 透明的；晶体的　　　　　　　　　　　　　　U3TB

cuprite /ˈkjuːpraɪt/ n. [矿物]赤铜矿　　　　　　　　　　　　　　U5TB

cylindrical /səˈlɪndrɪkl/ adj. 圆柱形的；圆筒状的　　　　　　　　U2TA

D

deleterious /ˌdeliˈtɪəriəs/ adj. 有害的；造成伤害的；损害的　　　　U1TA

density /ˈdensɪti/ n. 密度（固体、液体或气体单位体积的质量）　　　U5TA

deoxidize /diːˈɒksɪdaɪz/ v. 使（金属等）脱氧　　　　　　　　　　U3TA

deposition /ˌdepəˈzɪʃn/ n. 沉积（物）；沉淀（物）　　　　　　　　U3TB

desolate /ˈdesələt/ adj. 无人居住的；荒无人烟的；荒凉的　　　　　U6TB

detrimental /ˌdetrɪˈmentl/ adj. 有害的；不利的　　　　　　　　　U3TA

diameter /daɪˈæmɪtə(r)/ n. 直径　　　　　　　　　　　　　　　U2TA

die /daɪ/ n. 模具；冲模；压模　　　　　　　　　　　　　　　　　U2TA

diffuse /dɪˈfjuːz/ v. （使气体或液体）扩散，弥漫，渗透　　　　　　U3TB

digital twin 数字孪生　　　　　　　　　　　　　　　　　　　　U6TA

dismantle /dɪsˈmæntl/ v. 拆开，拆卸（机器或结构）　　　　　　　U6TB

dispose of 去掉；清除；销毁　　　　　　　　　　　　　　　　　　U5TB

dissolve /dɪˈzɒlv/ v. 使（固体）溶解　　　　　　　　　　　　　　U3TA

distortion /dɪˈstɔːʃn/ n. 变形；失真　　　　　　　　　　　　　　U5TA

dolomite /ˈdɒləmaɪt/ n. 白云石　　　　　　　　　　　　　　　　U3TB

domed /dəʊmd/ adj. 圆顶状的；半球形的　　　　　　　　　　　　U1TB

Glossary

dot /dɒt/ *v.* 星罗棋布于；遍布 　　　　　　　　　　　　　　　　　　U6TB

drill /drɪl/ *v.* 钻(孔)；打(眼) 　　　　　　　　　　　　　　　　　U2TA

ductility /dʌk'tɪləti/ *n.* 延展性；柔软性；顺从 　　　　　　　　　　　U4TA

E

effluent /'efluənt/ *n.* 流出物，流出液(尤指工厂排出的化学废料) 　　U5TB

elaboration /ɪˌlæbə'reɪʃn/ *n.* 详尽阐述 　　　　　　　　　　　　　U3TB

electric arc furnace (EAF) 电弧炉 　　　　　　　　　　　　　　　　U6TA

electric current assisted sintering 电流辅助烧结 　　　　　　　　　　U5TA

electrode /ɪ'lektrəʊd/ *n.* 电极 　　　　　　　　　　　　　　　　　U4TB

electrolysis /ɪˌlek'trɒləsɪs/ *n.* 电解 　　　　　　　　　　　　　　　U6TA

electrolytic /ɪˌlektrəʊ'lɪtɪk/ *adj.* 电解质的 　　　　　　　　　　　　U6TA

electrorefine /elektrɔːrɪ'faɪn/ *v.* 电解提纯；电精炼 　　　　　　　　U5TB

electrowinning /ɪ'lektrəʊˌwɪnɪŋ/ *n.* 电解冶金法；电解沉积 　　　　U4TB

empirical /ɪm'pɪrɪkl/ *adj.* 以实验(或经验)为依据的；经验主义的 　　U6TA

encompass /ɪn'kʌmpəs/ *v.* 包括，涉及(大量事物) 　　　　　　　　U3TA

extract /ɪk'strækt/ *v.* 提取；提炼 　　　　　　　　　　　　　　　　U1TB

extrusion /ɪk'struːʒn/ *n.* 挤压 　　　　　　　　　　　　　　　　　U2TA

F

fabricate /'fæbrɪkeɪt/ *v.* 制造；装配；组装 　　　　　　　　　　　　U5TA

fabrication /ˌfæbrɪ'keɪʃn/ *n.* 制造；装配；组装 　　　　　　　　　　U2TA

favourable /'feɪvərəbl/ *adj.* 有利的；有助于……的 　　　　　　　　U6TA

feed /fiːd/ *v.* 把……放进机器；将……塞进机器 　　　　　　　　　U5TB

feedstock /'fiːdstɒk/ *n.* 原料；给料 　　　　　　　　　　　　　　　U6TA

ferritic /fə'rɪtɪk/ *adj.* 铁素体的，铁氧体的 　　　　　　　　　　　　　U3TA

filament /ˈfɪləmənt/ *n.* （电灯泡的）灯丝；丝极　　　　　　　　　　　　　　U5TA

finely /ˈfaɪnlɪ/ *adj.* 成颗粒；细微地；细小地　　　　　　　　　　　　　　U4TB

flask /flɑːsk/ *n.* 烧瓶　　　　　　　　　　　　　　　　　　　　　　　　U6TB

flotation /fləʊˈteɪʃən/ *n.* 浮；漂浮　　　　　　　　　　　　　　　　　　U6TA

flux /flʌks/ *n.* 通量；流动　　　　　　　　　　　　　　　　　　　　　　U3TB

focal /ˈfəʊkl/ *adj.* 中心的；很重要的；焦点的；有焦点的　　　　　　　　U5TB

fold /fəʊld/ *v.* 折小，叠平　　　　　　　　　　　　　　　　　　　　　　U2TA

forge /fɔːdʒ/ *v.* 锻造；制作　　　　　　　　　　　　　　　　　　　　　U2TB

formability /ˌfɔːməˈbɪlɪtɪ/ *n.* 成型性，成型性能　　　　　　　　　　　　U3TA

fracture /ˈfræktʃə(r)/ *v.* （使）断裂，折断，破裂　　　　　　　　　　　U2TA

functionally graded materials 功能梯度材料　　　　　　　　　　　　　　U5TA

fuzzy /ˈfʌzɪ/ *adj.* （形状或声音）模糊不清的　　　　　　　　　　　　　U6TA

G

gabbro /ˈgæbrəʊ/ *n.* 辉长岩　　　　　　　　　　　　　　　　　　　　　U2TB

gangue /gæŋ/ *n.* 脉石　　　　　　　　　　　　　　　　　　　　　　　　U3TB

gaseous /ˈgæsɪəs/ *adj.* 似气体的；含气体的　　　　　　　　　　　　　　U3TA

geometry /dʒɪˈɒmɪtrɪ/ *n.* 几何形状；几何图形；几何结构　　　　　　　U5TA

glassy /ˈglɑːsi/ *adj.* 光亮透明的　　　　　　　　　　　　　　　　　　　U2TB

goggles /ˈgɒglz/ *n.* 护目镜；风镜；游泳镜　　　　　　　　　　　　　　U6TB

gravimetric /ˌgrævɪˈmetrɪk/ *adj.* （测定）重量的；重量分析的　　　　　U4TB

grind /graɪnd/ *v.* 使锋利；磨快；磨光　　　　　　　　　　　　　　　　U4TA

groundbreaking /ˈgraʊndbreɪkɪŋ/ *adj.* 开创性的；创新的；革新的　　　U5TA

gypsum /ˈdʒɪpsəm/ *n.* 石膏　　　　　　　　　　　　　　　　　　　　　U5TB

H

Glossary

hammer /ˈhæmə(r)/ *v.* (用锤子)敲，锤打　　　　　　　　　　　　　　　　U1TB

handful /ˈhændfʊl/ *adj.* 少数人(或物)　　　　　　　　　　　　　　　　U6TB

harden /ˈhɑːdn/ *v.* (使)变硬，硬化　　　　　　　　　　　　　　　　　U2TA

hollow /ˈhɒləʊ/ *adj.* 中空的；空心的　　　　　　　　　　　　　　　　U2TA

hollowware /ˈhɒləʊˌweə/ *n.* (由金属、瓷器等制成的)空心制品　　　　　U2TB

homologous /hɔˈmɔləgəs/ *adj.* （位置、结构等）相应的，类似的；同源的　U5TA

hot isostatic pressing 热等静压；高温等静力压制　　　　　　　　　　　U5TA

hydraulic /haɪˈdrɒlɪk/ *adj.* 液压驱动的　　　　　　　　　　　　　　　U2TA

hydrogen /ˈhaɪdrədʒən/ *n.* 氢；氢气　　　　　　　　　　　　　　　　U3TA

hydrometallurgical /ˈhaɪdrəʊˌmetəˈlɜːdʒɪkəl/ *adj.* 湿法冶金学的；水冶的　U5TB

I

identify /aɪˈdentɪfaɪ/ *v.* 确认；认出；鉴定　　　　　　　　　　　　　U4TA

implement /ˈɪmplɪmnt/ *v.* 使生效；贯彻；执行；实施　　　　　　　　　U6TA

impurity /ɪmˈpjʊərəti/ *n.* 杂质　　　　　　　　　　　　　　　　　　U1TB

in situ 在原位；在原地；在合适地方　　　　　　　　　　　　　　　　　U5TB

inasmuch as 因为；由于　　　　　　　　　　　　　　　　　　　　　　U5TB

incentive /ɪnˈsentɪv/ *n.* 激励；刺激　　　　　　　　　　　　　　　　U1TA

incl. 包括；连同……在内　　　　　　　　　　　　　　　　　　　　　　U6TA

indentation /ˌɪndenˈteɪʃn/ *n.* 凹陷　　　　　　　　　　　　　　　　U2TA

ingot /ˈɪŋgət/ *n.* (尤指金、银的)铸块，锭　　　　　　　　　　　　　　U1TA

installation /ˌɪnstəˈleɪʃn/ *n.* 安装的设备(或机器)；装置　　　　　　　U6TB

insulation /ˌɪnsjuˈleɪʃn/ *n.* 隔热；隔音；绝缘　　　　　　　　　　　　U3TB

integral /ˈɪntɪgrəl/ *adj.* 必需的；不可或缺的　　　　　　　　　　　　　U6TA

intermediate /ˌɪntəˈmiːdɪət/ *v.* 中级的；中等的；适合中等程度者的　　　U6TA

intervention /ˌɪntəˈvenʃn/ *n.* 出面；介入   U4TB

intricate /ˈɪntrɪkət/ *adj.* 错综复杂的   U5TA

inventory /ˈɪnvəntri/ *n.* (建筑物里的物品、家具等的)清单；财产清单   U5TB

IoT 物联网   U6TA

L

leachate /ˈliːtʃeɪt/ *n.* 沥滤物，沥滤液   U4TB

leach-solvent extraction 浸出溶剂萃取   U5TB

lead /liːd/ *n.* 铅   U6TA

lifecycle /ˈlaɪfˌsaɪkl/ *n.* 生命周期，寿命(产品等从开发到使用完毕的一段时间)   U5TA

limestone /ˈlaɪmstəʊn/ *n.* 石灰岩   U1TA

liquefy /ˈlɪkwɪfaɪ/ *v.* (使)液化   U5TA

loam /ləʊm/ *n.* 壤土；肥土   U2TB

low-alloy /ˈləʊˌælɔɪ/ *adj.* 低合金的   U3TA

lubricant /ˈluːbrɪkənt/ *n.* 润滑剂；润滑油   U5TA

M

machine vision (MV) 机器视觉   U6TA

maintenance /ˈmeɪntənəns/ *n.* 维护；保养   U6TA

malleable /ˈmæliəbl/ *adj.* 可锻造的；易成型的   U1TB

manganese /ˈmæŋɡəniːz/ *n.* 锰   U1TA

martensitic /ˌmɑːtinˈzitik/ *adj.* 马氏体的   U3TA

matte /mæt/ *n.* 锍   U6TA

melting point /ˈmeltɪŋ pɔɪnt/ 熔点   U1TA

metal injection molding 金属粉末注射成型   U5TA

metallurgist /məˈtælədʒɪst/ *n.* 冶金学家   U1TA

Glossary

microbe /ˈmaɪkrəʊb/ n. 微生物   U6TB

micron /ˈmaɪkrɒn/ n. 微米   U5TA

microorganism /ˌmaɪkrəʊˈɔːɡənɪzəm/ n. 微生物   U6TB

microsecond /ˈmaɪkrəʊˌsekənd/ n. 微秒   U5TA

mill /mɪl/ v. (用磨粉机)碾碎，磨成粉   U2TA

mold /məʊld/ n. 模具；铸模   U1TA

molten /ˈməʊltən/ adj. (金属、岩石或玻璃)熔化的；熔融的   U1TA

molybdenum /məˈlɪbdənəm/ n. 钼   U5TA

monetization /ˌmʌnɪtaɪˈzeɪʃn/ n. 货币化   U6TA

mortar /ˈmɔːtə(r)/ n. 灰泥；砂浆   U2TB

mottled /ˈmɒtld/ adj. 斑驳的；杂色的   U2TB

multitude /ˈmʌltɪtjuːd/ n. 众多；大量   U3TA

Neolithic /ˌniːəˈlɪθɪk/ adj. 新石器时代的   U1TB

N

nitrogen /ˈnaɪtrədʒən/ n. 氮；氮气   U3TA

normalize /ˈnɔːməlaɪz/ v. (使)正常化，标准化   U3TA

O

omnipresent /ˌɒmnɪˈprezənt/ adj. 无所不在的；遍及各处的   U4TB

open-hearth /ˌəʊpənˈhɑːθ/ adj. 使用平炉(炼钢)的   U1TA

optimal /ˈɒptɪməl/ adj. 最佳的；最适宜的   U6TA

optimization /ˌɒptɪmaɪˈzeɪʃən/ n. 优化   U6TA

ore /ɔː(r)/ n. 矿石   U1TA

ornament /ˈɔːnəmənt/ n. 首饰；饰物   U1TB

outskirts /ˈaʊtˌskɜːts/ n. (市镇的)边缘地带；市郊   U6TB

oversee /ˌəʊvəˈsiː/ *v.* 监督；监视   U2TB

oxidation /ˌɒksɪˈdeɪʃn/ *n.* 氧化   U3TA

oxidize /ˈɒksɪdaɪz/ *v.* (使)氧化；(尤指使)生锈   U2TB

P

parameter /pəˈræmɪtə/ *n.* 决定因素；规范；范围   U6TA

part /pɑːt/ *n.* 部件；零件   U5TA

patent /ˈpætnt/ *v.* 获得专利权   U1TB

pathogenic /ˌpæθəˈdʒenɪk/ *adj.* 致病的；病原的；发病的   U6TB

patina /ˈpætɪnə/ *n.* (金属表面的)绿锈，铜锈，氧化层   U5TB

patinize /ˈpætɪˌnaɪz/ *v.* 生绿锈   U5TB

payload /ˈpeɪləʊd/ *n.* (车辆等的)装载货物；装载量   U4TA

per annum /pər ˈænəm/ *adv.* 每年   U6TA

personnel /ˌpɜːsəˈnel/ *n.* (组织或军队中的)全体人员，职员   U6TA

pharmaceutical /ˌfɑːməˈsuːtɪkl/ *n.* 药物   U5TA

phase /feɪz/ *n.* 阶段；时期   U3TB

phosphorus /ˈfɒsfərəs/ *n.* 磷   U1TA

pig iron /ˈpɪɡ ˈaɪən/ *n.* 生铁；铸铁   U1TA

piglet /ˈpɪɡlət/ *n.* 猪仔；小猪   U1TA

pin /pɪn/ *n.* (插头的)销   U4TA

pinch /pɪntʃ/ *v.* 捏住；夹紧   U2TA

plague /pleɪɡ/ *n.* 瘟疫   U2TB

plant /plɑːnt/ *n.* 发电厂；工厂   U6TA

plasma arc cutting 等离子弧切割   U2TA

polymer /ˈpɒlɪmə(r)/ *n.* 聚合物；多聚体   U5TA

Glossary

porous /ˈpɔːrəs/ *adj.* 多孔的；透水的；透气的 — U5TA

powder forging [机] 粉末锻造 — U5TA

powered /ˈpaʊəd/ *adj.* 由……驱动的；电动的 — U1TB

precipitant /prɪˈsɪpɪtənt/ *n.* 沉淀剂 — U4TB

precipitation /prɪˌsɪpɪˈteɪʃn/ *n.* 沉淀；淀析 — U4TA

predetermined /ˌpriːdɪˈtɜːmɪnd/ *adj.* 预先确定的 — U3TA

predict /prɪˈdɪkt/ *v.* 预言；预告；预报 — U4TA

pregnant liquor 母液 — U5TB

preheat /ˌpriːˈhiːt/ *v.* 预热 — U1TB

primarily /ˈpraɪmərɪlɪ/ *adj.* 主要地；根本地 — U4TB

prospect /ˈprɒspekt/ *n.* 成功的机会；前景；前途 — U6TA

prototype /ˈprəʊtətaɪp/ *n.* 原型；雏形；最初形态；样品；模型 — U5TA

puddle /ˈpʌdl/ *n.* 水洼；小水坑 — U1TA

puddling process 搅炼法 — U1TB

punch /pʌntʃ/ *v.* （用打孔器等）打孔 — U2TA

pyrite /ˈpaɪraɪt/ *n.* [矿物] 黄铁矿 — U5TB

pyrometallurgical /paɪrɒmɪtæˈlɜːdʒɪkəl/ *adj.* 火法冶金的 — U5TB

Q

quench /kwentʃ/ *v.* 扑灭；熄灭 — U3TA

R

radiator /ˈreɪdɪeɪtə/ *n.* （车辆或飞行发动机的）冷却器，水箱 — U5TB

radius /ˈreɪdɪəs/ *n.* 半径（长度） — U2TA

raging /ˈreɪdʒɪŋ/ *adj.* 极其强大的；猛烈的 — U6TB

ram /ræm/ *n.* 撞击装置 — U2TA

ravage /ˈrævɪdʒ/ *n.* 毁坏；损坏　　　　　　　　　　　　　　　　　　　　U2TB

recrystallization /riːˌkrɪstəlaɪˈzeɪʃən/ *n.* 再结晶　　　　　　　　　　　U2TA

regenerator /rɪˈdʒenəreɪtə(r)/ *n.* 再生器　　　　　　　　　　　　　　　U1TB

reject /rɪˈdʒekt/ *n.* 废品；次品　　　　　　　　　　　　　　　　　　　U6TA

render /ˈrendə(r)/ *v.* 使成为；使处于某状态　　　　　　　　　　　　　U3TB

residual /rɪˈzɪdjuəl/ *n.* 残留物；剩余　　　　　　　　　　　　　　　　U3TA

retain /rɪˈteɪn/ *v.* 保持；持有；保留；继续拥有　　　　　　　　　　　U4TA

reutilization /riːjuːtəlaɪˈzeɪʃn/ *n.* 二次利用；重复利用　　　　　　　U3TB

reverberatory furnace 反射炉；反照炉　　　　　　　　　　　　　　　　U1TB

ribbon /ˈrɪbən/ *n.* 带状物；狭长的东西　　　　　　　　　　　　　　　U4TA

S

saw /sɔː/ *n.* 锯　　　　　　　　　　　　　　　　　　　　　　　　　　U2TA

scare /skeəs/ *adj.* 缺乏的；不足的；稀少的　　　　　　　　　　　　　U6TB

scrutinize /ˈskruːtnaɪz/ *v.* 仔细查看；认真检查；细致审查　　　　　　U4TB

seasoned /ˈsiːznd/ *adj.* 风干的，晾干的（可加工使用）　　　　　　　　U2TB

sedimentary /ˌsedɪˈmentri/ *adj.* 沉积的；沉积形成的　　　　　　　　　U2TB

semikilled /ˈseməkɪld/ *adj.* 半镇静钢的（半脱氧钢的）　　　　　　　　U3TA

shred /ʃred/ *v.* 切碎；撕碎　　　　　　　　　　　　　　　　　　　　　U4TB

silicate /ˈsɪlɪkeɪt/ *n.* 硅酸盐　　　　　　　　　　　　　　　　　　　　U2TB

silicon /ˈsɪlɪkən/ *n.* 硅　　　　　　　　　　　　　　　　　　　　　　　U1TB

simultaneously /ˌsɪmɪˈteɪniəsli/ *adv.* 同时发生（或进行）地；同步地　U5TA

sinter /ˈsɪntə/ *v.* （使）烧结；（使）熔结　　　　　　　　　　　　　　U5TA

slag /slæɡ/ *n.* 矿渣；熔渣；炉渣　　　　　　　　　　　　　　　　　　U1TB

slightly /ˈslaɪtli/ *adv.* 略微；稍微　　　　　　　　　　　　　　　　　　U6TA

Glossary

smelter /ˈsmeltə/ n. 熔炉	U5TB
solidification /səˌlɪdɪfɪˈkeɪʃn/ n. 凝固；浓缩	U3TA
solidify /səˈlɪdɪfaɪ/ v. (使)凝固，变结实	U2TB
solutionizing temperature 固溶温度	U4TA
sophisticated /səˈfɪstɪkeɪtɪd/ adj. 复杂巧妙的；先进的；精密的	U4TB
spectrum /ˈspektrəm/ n. 范围；各层次；系列	U6TA
spiegeleisen /ˈspiːɡ(ə)lˌaɪz(ə)n/ n. 镜铁	U1TA
stamp /stæmp/ v. 冲压	U2TA
stamp machine 压印机；冲床；烫金机；锤击机	U4TA
stiffness /ˈstɪfnɪs/ n. 不易弯曲；坚硬	U5TA
strand /strænd/ n. (线、绳、金属线等的)股，缕	U3TA
stride /straɪd/ n. 进展；进步；发展	U5TB
strip /strɪp/ v. 除去，剥去(一层)	U5TB
subdivision /ˈsʌbdɪvɪʒn/ n. 进一步分成的部分；分支；分部	U6TA
submerge /səbˈmɜːdʒ/ v. (使)潜入水中，没入水中，浸没，淹没	U4TB
substantially /səbˈstænʃəli/ adv. 非常；大大地	U1TB
substitute /ˈsʌbstɪtjuːt/ v. (以……)代替；取代	U6TA
suckling /ˈsʌklɪŋ/ n. 乳儿；乳兽	U2TB
sufficient /səˈfɪʃnt/ adj. 足够的；充足的	U6TA
sulfide /ˈsʌlfaɪd/ n. 硫化物	U3TB
sulfur /ˈsʌlfə(r)/ n. 硫	U3TA
sulfuric acid n. 硫酸	U5TB
supersonic /ˌsuːpəˈsɒnɪk/ adj. 超声速的	U1TB
supervisory /ˌsuːpəˈvaɪzəri/ adj. 监督的，指导的	U6TA

T

tailing /'teɪlɪŋ/ *n.* 残渣；尾料 — U5TB

take into account 考虑到；顾及 — U6TA

talcum /'tælkəm/ *n.* ［矿物］滑石 — U5TB

tapered /'teɪpəd/ *adj.* 锥形的 — U2TA

tectonic /tek'tɒnɪk/ *adj.* 地壳构造的 — U6TB

temper /'tempə(r)/ *v.* 使（金属）回火 — U3TA

tensile /'tensaɪl/ *adj.* 张力的；拉力的；抗张的 — U2TA

thermal /'θɜːml/ *adj.* 热的；热量的 — U3TA

thiobacillus /θaɪəubə'siləs/ *n.* 硫杆菌；产硫酸杆菌 — U6TB

titanium /tɪ'teɪnɪəm/ *n.*（symbol Ti）钛 — U4TA

trace /treɪs/ *n.* 微量；少许 — U4TB

trench /trentʃ/ *n.* 沟；渠 — U2TB

tungsten /'tʌŋstən/ *n.* 钨 — U1TB

tungsten carbide 硬质合金，［无化］碳化钨 — U5TA

tuyere /twiː'jɛr/ *n.* 鼓风口；风管嘴 — U1TB

U

upscale /'ʌpskeɪl/ *v.* 改进；扩大；升级 — U6TA

uranium /jʊ'reɪnɪəm/ *n.* 铀（放射性化学元素） — U6TB

V

venture /'ventʃə/ *n.*（尤指有风险的）企业，商业，经营项目 — U6TB

versatile /'vɜːsətaɪl/ *adj.* 多用途的；多功能的 — U1TA

vessel /'vesl/ *n.* 容器 — U2TA

viscosity /vɪ'skɒsəti/ *n.* 黏性；黏度 — U3TB

Glossary

vitreous /ˈvɪtriəs/ *adj.* 玻璃质的；透明的 　　　　　　　　　　　　　　　　　U3TB

W

waterfront /ˈwɔːtəfrʌnt/ *n.* 滨水区；码头区 　　　　　　　　　　　　　　　U2TB

waterjet /ˈwɔːtədʒet/ *n.* 喷水 　　　　　　　　　　　　　　　　　　　　　U2TA

wear out 磨薄；穿破；磨损；用坏 　　　　　　　　　　　　　　　　　　　　U4TA

weathering /ˈweðərɪŋ/ *n.* （岩石的）风化 　　　　　　　　　　　　　　　　U5TB

weld /weld/ *v.* 焊接；熔接 　　　　　　　　　　　　　　　　　　　　　　　U2TA

workpiece /ˈwɜːkpiːs/ *n.* 工作部件；工件 　　　　　　　　　　　　　　　　U2TA

Z

zinc /zɪŋk/ *n.* 锌 　　　　　　　　　　　　　　　　　　　　　　　　　　　U6TA